Studies in Linguistics

Edited by
Laurence Horn
Yale University

Studies in Linguistics

LAURENCE HORN, *General Editor*

Lenition and Contrast
The Functional Consequences of Certain Phonetically Conditioned Sound Changes
Naomi Gurevich

The Inflected Infinitive in Romance Languages
Emily Scida

Syntactic Form and Discourse Function in Natural Language Generation
Cassandre Creswell

The Syntax-Information Structure Interface
Evidence from Spanish and English
Eugenia Casielles-Suárez

The Ups and Downs of Child Language
Experimental Studies on Children's Knowledge of Entailment Relationships and Polarity Phenomena
Andrea Gualmini

Category Neutrality
A Type-Logical Investigation
Neal Whitman

Markedness and Faithfulness in Vowel Systems
Viola Miglio

Phonological Augmentation in Prominent Positions
Jennifer L. Smith

Enriched Composition and Inference in the Argument Structure of Chinese
Ren Zhang

Discourse Adjectives
Gina Taranto

Structural Markedness and Syntactic Structure
A Study of Word Order and the Left Periphery in Mexican Spanish
Rodrigo Gutiérrez-Bravo

Negation and Licensing of Negative Polarity Items in Hindi Syntax
Rajesh Kumar

Syllable Weight
Phonetics, Phonology, Typology
Matthew Kelly Gordon

Reassessing the Role of the Syllable in Italian Phonology
An Experimental Study of Consonant Cluster Syllabification, Definite Article Allomorphy and Segment Duration
Kristie McCrary Kambourakis

Telicity and Durativity
A Study of Aspect in Dëne Suliné (Chipewyan) and German
Andrea Wilhelm

Artificial Hearing, Natural Speech
Cochlear Implants, Speech Production, and the Expectations of a High-Tech Society
Joanna H. Lowenstein

Artificial Hearing, Natural Speech

Cochlear Implants, Speech Production, and the Expectations of a High-Tech Society

Joanna H. Lowenstein

Routledge
Taylor & Francis Group
270 Madison Avenue
New York, NY 10016

Routledge
Taylor & Francis Group
2 Park Square
Milton Park, Abingdon
Oxon OX14 4RN

Routledge is an imprint of Taylor & Francis Group, an Informa business

International Standard Book Number-13: 978-0-415-97604-6 (Hardcover)

Visit the Taylor & Francis Web site at
http://www.taylorandfrancis.com

and the Routledge Web site at
http://www.routledge.com

Contents

Tables

Figures

Acknowledgments

This dissertation would not have been possible without the support of Mario Svirsky at Indiana University through his NIH-NIDCD grant R01-DC03937, subject funding from Dean Janel Mueller, the support of the Language Laboratories and Archives at the University of Chicago, or the participation of my subjects MS, AM, and DS.

It is my greatest regret that Karen Landahl was unable to see this project to completion. From the days of my earliest thoughts on cochlear implants and modern culture, she was incredibly supportive and encouraging. Her influence is present throughout this dissertation.

Thanks to my committee, Bill Darden, Susan Goldin-Meadow, and Sean Fulop. Bill's willingness to take over as chair and Sean's willingness to join my committee were bright lights in a very dark February.

Thanks to my parents, Claudia and Carl Lowenstein, for their support during this long journey. The calculation scripts that my father wrote eased the path considerably. Special thanks also must go to Isabel Guzman-Barron, the best roommate ever, and to my closest friends: Jen Nissen, Jason Frankel, Beth Kiedrowicz, Lisa Rogers Lowrance, Nathan Stohlmann, Lisa Cohen, Betsy Lundsten, Gar Francis, and Tim Pierce. My voice teacher, Marina Gilman, proved an invaluable sounding board over the years. Thanks also to the Wombats, the denizens of a.p, and the LiveJournal crowd.

And thanks to the members of ChiPhon and the phonetics reading group over the years, especially Tami Wysocki, Greg Davidson, Elisa Steinberg, Rebekka Egger, Yi Xu, Yukari Hirata, Jeannette Denton, Rachel Hemphill, Audra Dainora, Alex Francis, and Peter Viechnicki. Lec Maj and Barbara Need also provided technical support and advice.

The revision of this dissertation into book form was supported by Grant No. R01 DC 00633 from the National Institute on Deafness and Other Communication Disorders to Susan Nittrouer.

1. INTRODUCTION

"Better. Stronger. Faster." These descriptions of the Six Million Dollar Man and the Bionic Woman, recited before the opening credits of every episode, reflect the hopes of modern society that technology, especially computer technology, can fix or even enhance the human body, damaged or undamaged. From human interest news stories to science fiction series, technology is presented as a way to overcome the frailties of the human body. Local and national news services report on scientists striving to find a cure for paralysis (by both biological and technological means); the hero of the 1994-95 Fox science fiction television series *M.A.N.T.I.S*, a man who is paralyzed from the waist down, becomes "superhuman" when he dons a complex powered suit. On the WB's *Birds of Prey* (Fall 2002), Oracle (formerly Batgirl) supervises and coordinates two other superheroes from her high-tech wheelchair command center. The "Spare Parts" segment of *Superhuman Body* on The Learning Channel (2001), which features high-tech prosthetic arms, hand transplants, cross-species brain cell transplants, and cochlear implants opens with the following quote: "In principle, it's possible to replace all the parts that make up a car. The idea of doing the same thing with the human body, as our parts wear out, or are damaged by disease, is seductive."

Cochlear implants (CIs), first envisioned in the 1960s and popularized in the late 1980s as a replacement for the damaged inner ear in cases of nerve deafness, continue in this trend of using technology to "fix" the damaged body. News coverage of cochlear implants, particularly stories about deaf children receiving implants, demonstrate that the concept of "bionic" hearing has caught the imagination of our society. These stories often gush about how wonderful it is that these children will be able to finally hear; even stories about adults receiving cochlear implants fall into a similar vein. Cochlear implants were featured in several documentary and dramatic presentations on U.S. television between 1998 and 2002. But do these documentaries and presentations reveal a realistic picture of the ordinary cochlear implantee? Could they lead to unrealistic expectations?

There are still many things we do not know about how language works. In particular, the complex relationships between perception and production in adults are still not completely understood. This sort of uncertainty contributes to the difficulty in evaluating whether a complex device like a cochlear implant lives up to all expectations. For example, should the term 'success' only be reserved for the implantees who can understand speech without visual input (considered the highest level of performance)? How should changes in production be evaluated?

These evaluation difficulties are in no small part related to the controversies over the implantation of very young deaf children. The Deaf cultural community objects to the "experimental" nature of the cochlear implant; the various definitions of "benefit" add to the confusion. According to Lane, Hoffmeister, and Bahan (1996):

> ...the less demanding the definition of 'benefit,' the greater the percent of implanted children who receive that benefit. Thus, nearly all implanted children can detect environmental sound. A large number can identify many of those sounds. Many implanted children can distinguish some properties of speech sounds even if they cannot identify those sounds. Some children achieve higher lipreading scores with an implant than without one and a few can understand spoken language...(397)

In contrast, a study published in *The Journal of the American Medical Association* in August of 2000 that calculated a cost-utility analysis of cochlear implant use found that "cochlear implants in profoundly deaf children have a positive effect on quality of life at reasonable direct costs and appear to result in a net savings to society" (Cheng, Rubin, Powe, Mellon, Francis, and Niparko 2000, 850). Clearly, the degree to which cochlear implants are perceived as acceptable or desirable depends on the way their "benefits" are defined. Even in the less controversial area of postlingually deafened adults with cochlear implants, it can be difficult to evaluate to what degree patients are aided by them.

Acoustic studies of the speech of postlingually deaf adults, primarily conducted by Joseph Perkell at MIT in the mid to late 1990s, provide evidence that analog cochlear implants allow a sufficient degree of auditory feedback to affect vowel and consonant articulation and voice pitch, generally resulting in more "normal" speech. This dissertation replicates and extends those studies by examining the speech of three postlingually deaf adults with modern digital cochlear implants, focusing on vowel formants and duration, voice onset time and syllable duration, and fundamental frequency. The postlingually deaf adults also participated in a perception task, and their speech was categorized by both naïve and expert listeners. In addition, three televised documentaries and episodes from three television shows that featured cochlear implant users were analyzed for accuracy of information presented, and for what expectations about CI users might arise from viewing them.

2. BACKGROUND

2.1 THE EAR: STRUCTURE AND FUNCTION

The normally functioning ear is usually described in three sections: the outer ear, middle ear, and inner ear (cochlea).

(1) The outer ear. The external portion of the ear consists of the pinna, concha, external auditory canal, and tympanic membrane (eardrum). The outer ear serves to protect the eardrum and to collect sound. The resonance of the concha is approximately 5 kHz and the resonance of the external auditory canal is approximately 2.5 kHz. These resonances complement each other to amplify the 1500 Hz – 7000 Hz frequency range 10 to 15 decibels (Yost 2000, 71). The pressure of the sound wave sets the tympanic membrane into motion, which transfers the sound to the middle ear.

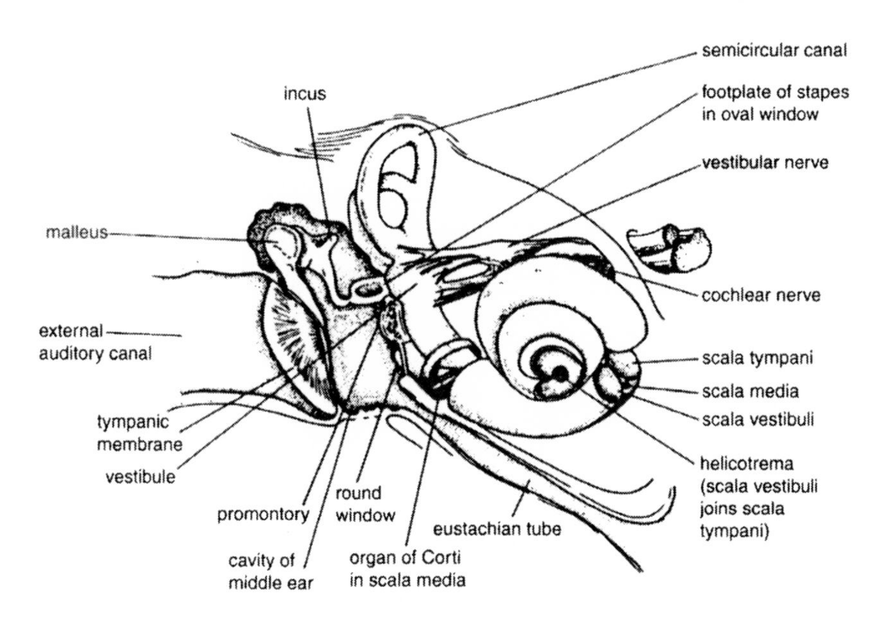

Figure 2.1 Structure of the inner ear

(2) The middle ear. The middle ear acts to transfer the motion of the tympanic membrane to the fluid of the cochlea. Inner ear fluid is significantly denser than air, so the pressure levels at the eardrum would not be sufficient to move the fluid in the cochlea. The ossicular chain (malleus, incus, and stapes) transfers and magnifies the vibrations from the eardrum to the oval window of the cochlea.

(3) The inner ear. The inner ear consists of the vestibular system (semicircular canals) and the cochlea. The cochlea acts to transfer the fluid motion within the cochlea to neural impulses in the auditory nerve (VIIIth cranial nerve). The oval window membrane vibrates, setting the fluids within the cochlea into motion. The fluid pressure wave travels along the scala vestibuli to the helicotrema (the apex of the cochlea) and back along the scala tympani, which terminates in the round window. The basilar membrane (approximately 35 mm in length (Yost 2000, 104)), which separates the scala vestibuli and scala media (cochlear duct), is displaced by the fluid pressure wave in a traveling wave from base to apex. How far towards the apex it travels depends on the frequency of stimulation; lower frequencies travel farther.

As illustrated in Figure 2.2, a 60 Hz sine wave travels the furthest towards the apex of the basilar membrane, while a 2000 Hz sine wave pri-

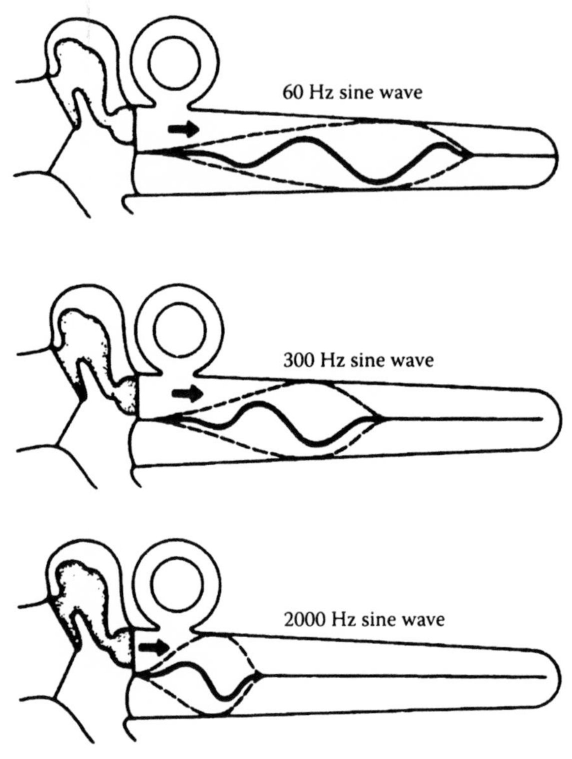

Figure 2.2 Illustrations of the fluid pressure wave

marily stimulates the basal end of the membrane. Note that the basal end of the basilar membrane responds to low frequency stimulation as well as high frequency stimulation, while the apical end responds only to low frequency stimulation (Yost 2000, 98). As the sound stimulus level increases, the degree of displacement in the basilar membrane increases, and "the temporal pattern of basilar membrane displacement follows that of the stimulating sound" (Yost 2000, 99). Taking these factors into account, it is clear that the movement of the basilar membrane "corresponds with the frequency, level, and temporal pattern of acoustic stimulation" (99). Any given location along the basilar membrane acts as a bandpass filter, as that location will be "displaced maximally by only one frequency" (99). The neural output of the cochlea reflects this phenomena, where the neural responses at the basal end of the cochlea correspond to high frequencies and the neural responses at the apical end correspond to low frequencies. This is often referred to as the *tonotopic* organization of the cochlea.

The cochlear duct, or scala media, runs between the scala vestibuli and the scala tympani, along the length of the cochlea, and contains the organ of Corti. The organ of Corti contains inner and outer hair cells. Displacement of the basilar membrane causes shearing forces on the hair cells in the organ of Corti (see Figure 2.3). The shearing forces activate the neurons at the base of the hair cells, which send neural discharges to the auditory nerve bundle (along the inner curve of the cochlea). Nerve fibers from the apex of the cochlea (low frequency information) are in the center of the bundle, and nerve fibers from the other turns of the cochlea (high frequency information) are on the outside (Yost 2000, 123).

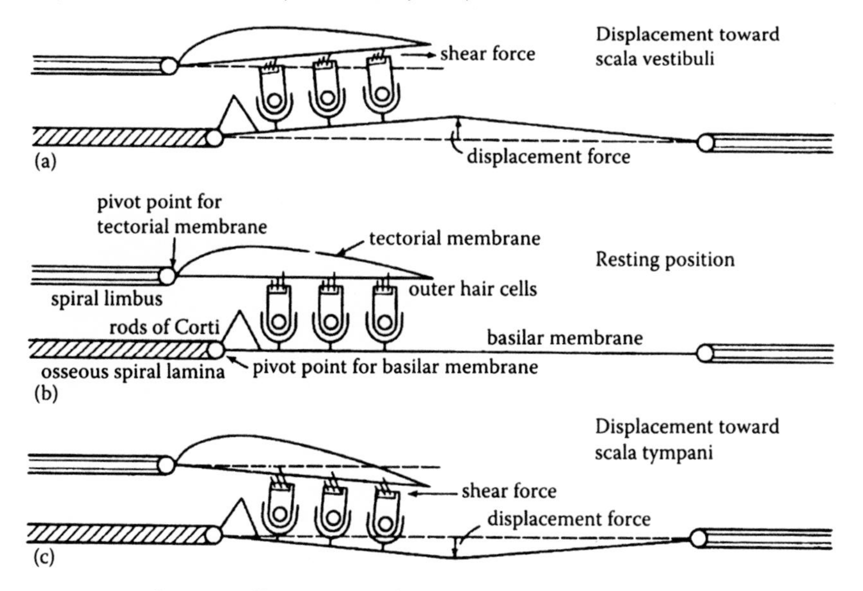

Figure 2.3 Schematic illustrations of the interaction between the basilar membrane and the organ of Corti

The auditory nerve bundle combines with nerves carrying signals from the vestibular system to make up the VIIIth cranial nerve (auditory nerve). The auditory nerve carries the signal output from the cochlea to the auditory processing centers of the brain.

2.2 COCHLEAR IMPLANTS

2.2.1 General structure and function

Cochlear implants, independent of manufacturer, can be divided up by function and location into three primary parts.

(1) Sound location (external): A microphone (which is generally located in the earpiece of the implant, but may also be worn on the body) collects sound and transmits it to the speech processor. This can be compared to the function of the outer and middle ear.

(2) Sound processing (external): A speech processor (which is approximately the size of a pager, worn on the body or integrated into the same earpiece as the microphone) filters the sound information collected by the microphone, performs acoustic analysis on the signal, and translates the results of that analysis into a coded signal. This can be compared to the function of the cochlea and sound processing center of the brain.

(3) Sound transmission (external and internal): The coded signal from the speech processor is transmitted via an externally worn transmitting coil to the cochlear implant, which converts the coded signal into appropriate electrical impulses along the array of electrodes inserted into the cochlea. This can be compared to the function of the cochlea.

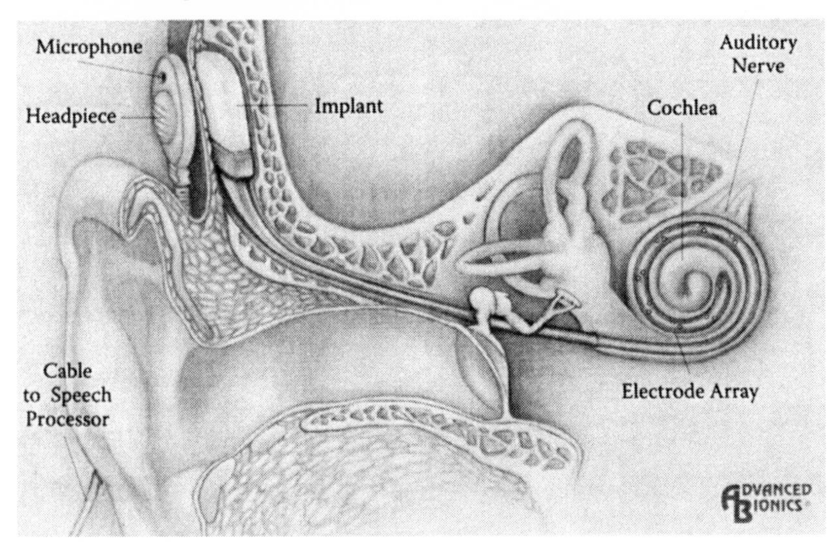

Figure 2.4 Illustration of the Clarion cochlear implant in situ

From the first experiments with multiple electrode arrays, cochlear implants have been designed to utilize the tonotopic organization of the cochlea. Cochlear implants work by directly stimulating the nerve endings in the cochlea, which then transmit the information to the sound processing centers of the brain.

Cochlear implant surgery takes two to three hours, and is conducted under general anesthesia. The appropriate area of the head is shaved, and the surgeon makes a curved incision behind the ear. A small depression is created in the mastoid bone to hold the cochlear implant in place. The surgeon then inserts the electrode array into the cochlea through an opening near the round window. It is important that the implant electrode be placed as deeply as possible, so that the tip of the electrode array can stimulate the appropriate area of the cochlea for low frequency sounds. The incision is closed and the head bandaged (Cochlear Corporation 1999, 24).

After the incision has healed (three to five weeks), the implantee meets with an audiologist to be fitted with the external portions of the implant (speech processor, microphone, etc.). Using a computer, the audiologist sends signals to each electrode, increasing the stimulation level until the recipient indicates that a sound is heard, and continuing until they indicate the maximum comfortable loudness level. This results in a speech processor program that stimulates each electrode appropriately, and takes approximately one hour. The audiologist sets up a series of follow up appointments so that the programming can be adjusted and fine-tuned, generally at one month, three months, six months, and one year (Sheryl Erenberg, audiologist; personal communication, 2001). The role of the audiologist is particularly crucial, as any given electrode may not line up with active nerve cells and thus will not respond to stimulation, and the remaining electrodes need to be programmed to compensate so that no part of the frequency range is left unstimulated. For example, suppose that the default setting of electrode A is activation for the 3000-3400 Hz range, electrode B for the 3400-3800 Hz range, and electrode C for the 3800-4200 Hz range. If electrode B does not respond, electrode A would be reprogrammed for the 3000-3600 Hz range and C for the 3600-4200 Hz range, making sure no part of the frequency range was omitted.

2.2.2 Cochlear implants: Structures and Processing Strategies

2.2.2.1 Ineraid

Current cochlear implants (Nucleus, Clarion, Med-El) all use digital processing strategies, but this has not always been the case. The 4-electrode, 4-channel Ineraid implant (Richards Medical Co.) uses an analog processing system. The body-worn speech processor for this implant utilizes four overlapping analog bandpass filters with center channel frequencies of 500 Hz, 1000 Hz, 2000 Hz, and 3400 Hz, and filter crossover points at approximately 700 Hz, 1400 Hz, and 2300 Hz. The analog output of each filter is transmitted to a separate intracochlear electrode (total of four). The most apical electrode is placed approximately 22 mm from the round window of the cochlea, and the electrodes are located 4 mm apart (Matthies, Svirsky, Perkell, and Lane 1996).

2.2.2.2 Nucleus

The Nucleus (Cochlear Corporation) multichannel cochlear implant started out as a 22-electrode implant (20 active electrodes). In recent years, they have released a 24-electrode implant (22 active electrodes, and two electrodes that allow for internal error checking). This uncurved electrode array is "inserted approximately one inch [25 mm] into the cochlea during surgery" (Cochlear Corporation 1999, 9). The speech processor accompanying the implant has gone through several incarnations over the history of the Nucleus implants. The first processor, WSP (Wearable Speech Processor) could be programmed with the F_0/F_2 or $F_0/F_1/F_2$ speech processing strategies. MSP (Mini Speech Processor), in clinical use since 1989, could be programmed with F_0/F_2, $F_0/F_1/F_2$, or MPEAK strategies (Skinner, Clark, Whitford, Seligman, Staller, Shipp, Shallop, Everingham, Menapace, Arndt, Antogenelli, Brimacombe, Pijl, Daniels, George, McDermott, and Beiter 1994, 16). WSP and MSP are both body worn processors.

The first strategy used, F_0/F_2, extracted the fundamental frequency (F0) and the frequency of peak sound energy in the 800-2300 Hz region (F2), divided into logarithmically equal pass-bands. The F2 information determined which electrode was stimulated (electrodes were assigned in an apical-to-basal order) and the amplitude of stimulation, and the F0 information determined the rate of stimulation. $F_0/F_1/F_2$ added information in the 300-1000 Hz range (F1) and the bandwidth for F2 information was also adjusted (1000-4000 Hz). F1 frequency bands were divided in equal linear steps, assigned to the apical third of electrodes, and F2 frequency bands were divided in equal logarithmic steps and assigned to the remaining basal electrodes. Multipeak (MPEAK) added amplitude information for three high frequency bands (2000-2800 Hz, 2800-4000 Hz, 4000-6000 Hz), delivered to three fixed basal electrodes (Skinner et al. 1994, 16).

The most recent speech processors are SPrint (body worn) and ESPrit (ear level). The SPrint processor can be programmed with the SPEAK, CIS, and ACE speech processing strategies, and the ESPrit processor can be programmed with the SPEAK processing strategy (Cochlear Corporation 1999, p. 10-13).

The Spectral Peak (SPEAK) processing strategy operates in a very different fashion from the F_0/F_2, $F_0/F_1/F_2$, or MPEAK strategies. This processing strategy

> uses a bank of 20 bandpass filters, each allocated to a single electrode, to process the incoming signal over the range 150 to 10,823 Hz. A "spectral maxima detector" constantly monitors the energy within each filter band and selects those bands with the largest amplitudes. Those bands selected receive stimulation. The electrodes are stimulated sequentially from those located in the basal regions of the cochlea (higher frequencies) to apical regions (lower frequencies). The actual number of electrodes selected per stimulus cycle depends on the spectral composition and level of the incoming signal. On average, 6 electrodes are stimulated per cycle, with the actual number of any given cycle restricted to the range of 1 to 10...The rate of stimulation per electrode varies between about 180 and 300 pulses per second, dependent on the number of spectral maxima selected, the signal intensity, and the individual's program parameters (Parkinson, Parkinson, Tyler, Lowder, and Gantz 1998, 1076).

The average stimulation rate for SPEAK is 250 pulses per second (pps) per channel (Fourakis, Skinner, Holden, and Holden 2001, 32).

The F_0/F_2, $F_0/F_1/F_2$, and MPEAK strategies all extract linguistically relevant features from the acoustic signal, while SPEAK presents an overall picture of the acoustic signal. The following figure (see Figure 2.5), from Seligman and McDermott 1995, compares (A) a spectrogram of the word "choice" with (B) representations of the output of the SPEAK processing strategy and (C) the MPEAK processing strategy (labeled Mini Speech Processor output).

Continuous Interleaved Sampling (CIS) uses "nonsimultaneous interleaved pulses as stimuli, with amplitudes derived from the envelopes of bandpass filter outputs. Envelope signals are formed by rectification and lowpass filtering, and the amplitude of each stimulus pulse is determined by a nonlinear transformation of the corresponding channel's envelope signal at that time" (Wilson, Lawson, Finley and Wolford 1993, 374). Ziese, Stützel, von Specht, Begall, Freigang, Sroka, and Nopp 2000 describe CIS as follows: "Each channel of the multi-channel device is associated with a certain frequency range of the sound signal. All channels are stimulated in each cycle, each channel with an amplitude derived from the instantaneous energy content of the associated frequency range" (321). CIS is described as providing more detail than SPEAK for the timing of speech, as the "relatively high rates of pulsatile stimulation used in CIS processors allow representation of rapid variations in speech by rapid changes in pulse amplitude" (Wilson et al. 1993, 374). Programmed stimulation rates for CIS range from 720 to 2400 pps per channel, with the total stimulation rate limited to 14,400 pps/cycle across channels (Fourakis et al. 2001, 32).

Advanced Combination Encoders (ACE) "uses SPEAK's roving selection strategy (choosing from among 22 separate stimulation sites) combined

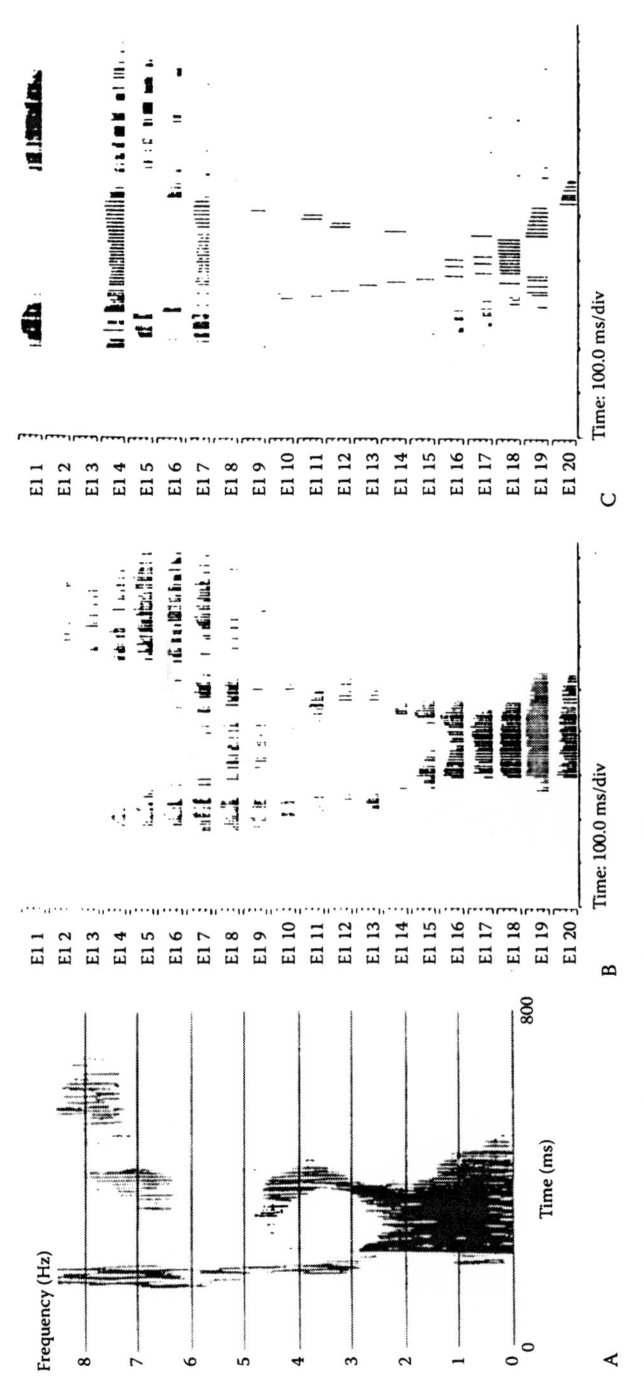

Figure 2.5 Illustration of MPEAK and SPEAK processing strategies

with CIS's option of higher stimulation rate" (Cochlear Corporation 1999, 13). Fourakis et al. (2001) describe ACE as follows: "Incoming sound is filtered into as many as 22 bandwidths. During each processing cycle, the signal in the specified number of bands [6-12] with the highest amplitude causes stimulation on the associated electrodes. The rate of stimulation can vary between 250 and 2400 pps/channel as long as the total stimulation rate across channels does not exceed 14,400 pps/cycle" (32).

2.2.2.3 Clarion

The Clarion (Advanced Bionics) cochlear implant is a 16-electrode, eight-channel implant (one pair of electrodes per channel), and recently (1999) a 16-channel implant (Platinum Series) has been approved by the FDA (*CII Bionic Ear Implant*). The Clarion electrode array is curved so as to maintain contact with the medial wall of the cochlea, and can be inserted to a maximum depth of 25 mm (Kessler 1999, 9). The 16-channel Platinum Series implant is available with two speech processors, the Platinum Sound Processor (body worn) and the Platinum BTE (ear level). These processors can be programmed with the SAS, PPS, CIS, and hybrid speech processing strategies (*Platinum Series Implant*).

The Clarion cochlear implant uses a different set of speech processing strategies from the Nucleus implant. Simultaneous Analog Stimulation (SAS) "is a fully simultaneous strategy that digitally reconstructs the analog waveform…." When 7 channels are operating simultaneously, it samples the input signal at a rate of 91,000 samples per second" (Zimmerman-Phillips and Murad 1999, 17). Paired Pulsatile Sampler (PPS), described as a "partially simultaneous strategy," stimulates two distant channels simultaneously, effectively doubling the stimulation rate (Kessler 1999, 15). The Clarion version of CIS (Continuous Interleaved Sampling) operates similarly to CIS in the Nucleus speech processors and "delivers pulsatile stimulation sequentially across the 8 channels at a rate of 6,500 pulses per second" (Zimmerman-Phillips and Murad, 17).

Hybrid speech processing strategies are currently under development, combining SAS and CIS, "with simultaneous analog stimulation applied in

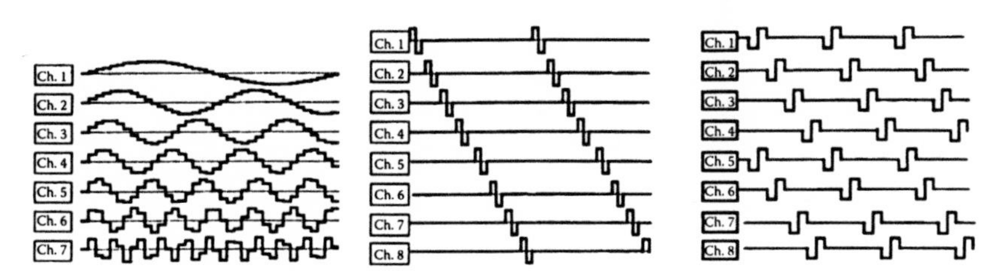

Figure 2.6 Illustration of the SAS, CIS, and PPS stimulation patterns

the lower frequencies and non-simultaneous pulsate stimulation in the high [frequencies]" (Kessler, 15).

2.2.2.4 *Med-El*

The Med-El Combi 40+ cochlear implant is currently undergoing clinical trials in the U.S. It contains 24 active electrodes (12 channels), and the electrode carrier is "specially designed to enable deep placement [30 mm]" (Medical Electronics 1999, 5). Two speech processors are available for the Combi 40+: the body worn CIS PRO+ and the ear level TEMPO+ (7). The CIS PRO+ processor can be programmed with the High-Rate CIS and High-Rate n-of-m speech processing strategies, while the TEMPO+ processor is programmed with the CIS+ processing strategy (8).

High-Rate CIS, which operates similarly to CIS in the Clarion and Nucleus implants, can operate over 7 or 12 channels and has a maximum sequential stimulation rate of 18,180 pulses per second (Ziese et al., 321). CIS+ is described as an enhanced version of CIS that uses the Hilbert transform for envelope extraction and "allows more accurate determination of the signal envelope containing loudness-over-time and pitch information" (Helms et al. 2001, 32). CIS+ also analyzes over a larger and higher frequency range than High-Rate CIS (200-10,000 Hz, as compared to 300-5,500 Hz) (33). High-Rate n-of-m is comparable to the Nucleus SPEAK strategy; "in each cycle only the *n* energy-richest channels out of the total number *m* of channels are stimulated based on a channel selection process the result of which is renewed from one cycle to the next" (Ziese et al. 322). High-Rate n of m uses a constant *n* number of channels (default is 7), which is adjustable by the audiologist, and differs from SPEAK where the number *n* of stimulated channels is derived from properties of the sound signal (322).

2.2.3 Cochlear implants: Implantees

Children with cochlear implants (CIs) and adults with cochlear implants comprise very distinct groups. Many children with CIs are prelingually deafened, that is, they were born deaf or became deaf before they were old enough to have acquired a spoken language. Since 1992, many are implanted as young as 7 months to 2 years old, though ages of prelingually deafened child implantees range into the mid-teens. Other children with CIs are considered postlingually deafened, with varying amounts of auditory and spoken language experience. Studies of prelingually deafened children with cochlear implants find that it takes at least two years of intense auditory rehabilitation to see results in perception and production tasks (Carpenter 1997)[1]. There is a far lower success rate with prelingually deafened adults; according to Waltzman and Cohen's 1999 review of studies of prelingually deafened adults with cochlear implants, "the prognosis for adolescents and adults with congenital and/or prelingual hearing losses attaining open-set speech recognition is considered to be poor" (84). The majority of adults who receive cochlear implants became deaf after they learned to speak. The speech of these postlingually deafened adults is described as having greater variations in pitch and a lesser degree of place differentiation in fricative and stop consonants (Lane and Webster 1991) and a more restricted vowel space and exaggerated pitch contrasts (Waldstein 1990) than comparable hearing subjects. Intelligibility studies of the speech of postlingually deafened adults (Cowie, Douglas-Cowie, and Kerr 1982, Goehl and Kaufman 1984) find the degree to which speech intelligibility is reduced varies greatly–some subjects maintain normal intelligibility and for others, intelligibility is impaired. These adults do not usually require extensive auditory rehabilitation, and often show marked perceptual improvement within six months.

2.3 RESEARCH ON ADULTS AND CHILDREN WITH COCHLEAR IMPLANTS

Much research that uses postlingually deafened adults with cochlear implants as their subjects focus on post-implant changes in perception (Dorman, Loizou, Spahr, and Maloff 2002; Friesen, Shannon, Baskent, and Wang 2001; Fu and Galvin 2001; Fu, Galvin, and Wang 2001; Geurts and Wouters 2001; Hamzavi, Baumgartner, and Pok 2001; Kirk, Tye-Murray, and Hurtig 1992; Loizou, Dorman, and Powell 1998; Loizou and Poroy 2001; McKay and Henshall 2002; Richardson, Busby, Blamey and Clark 1998; Svirsky, Silveira, Suarez, Neuberger, Lai, and Simmons 2001; Tyler & Moore 1992; Välimaa, Määttä, Löppönen, and Sorri 2002(1), 2002(2); Vandali 2001; van Hoesel, Tong, Hollow, and Clark 1993; Zeng, Grant, Niparko, Galvin, Shannon, Opie, and Segel 2001). This is also true in the literature on children with cochlear implants (Blamey, Sarant, Paatsch, Barry, Bow, Wales, Wright, Psarros, Rattigan, and Tooher 2001; Ciocca, Grancis, Aisha, and Wong 2002; Fryauf-Bertschy, Tyler, Kelsay, and Gantz 1992; Fryauf-Bertschy, Tyler, Kelsay, Gantz, and Woodworth 1997; Tye-Murray 1992).

Those papers that do deal with changes in production in children with cochlear implants primarily focus on expert hearer evaluation of subjects' speech (Tye-Murray, Spencer, and Gilbert-Bedia 1995; Tye-Murray, Spencer, and Woodworth 1995), on comparing close transcription of speech to the speakers' intent (Blamey, Barry, Bow, Sarant, Paatsch, and Wales 2001; Blamey, Barry, and Jacq 2001; Chin 2002; Geers, Spehar, and Sedey 2002; Ingram 2002; Osberger, Maso, and Sam 1993; Tye-Murray & Kirk 1993; Tye-Murray, Spencer, Gilbert Bedia, and Woodworth 1996), or on the ratings of naïve listeners (Chin, Finnegan, and Chung 2001). Geers 2002 used both expert transcribers and naïve listeners. I have encountered only three papers which analyze the speech of a child with a cochlear implant using acoustic analysis (Economou, Tartter, Chutte, and Hellman 1992; Ertmer 2001; Goffman, Ertmer, and Erdle 2002).

There has been very little research on the acoustics of speech produced by adult cochlear implant (CI) users. The primary acoustic research on the speech production of adult CI users consists of nine papers by researchers at MIT, seven of which were published between 1992 and 1997. These papers are primarily concerned with the role hearing plays in regulating speech production. Through acoustic comparison of speech recorded before and at regular intervals after implantation, the authors determine which aspects of speech production change after experience with a cochlear implant. The studies which use an experimental design where the implant is turned off for a period of time before the subjects are recorded, and then turned on and their speech recorded again, reveal the short-term effects of auditory feedback from the cochlear implant. I will describe these studies in detail, specifically in the areas of materials recorded and methods of instrumental analysis. These areas are important in regards to evaluating and replicating this research.

2.4 THE PERKELL GROUP STUDIES

2.4.1 The Ineraid cochlear implant

The following papers focus on the speech of postlingually deafened adults implanted with the 4-channel analog Ineraid implant (Richards Medical Co.). See **2.2.2.1** for details on this implant's electrode array and speech processing strategy.

2.4.2 Vowel production studies

Perkell, Lane, Svirsky, and Webster (1992) and Svirsky, Lane, Perkell, and Wozniak (1992) examine vowel production. In Perkell et al. 1992, subjects MA and MB (male), and FA and FB (female) were recorded twice before implant activation, and at five intervals afterwards: immediately at activation, and 4 weeks, 12 weeks, 26 weeks, and 52 weeks post-activation. The subjects were recorded speaking three repetitions each of nine vowels /i/, /ɪ/, /ɛ/, /æ/, /ɑ/, /ʌ/, /ɔ/, /ʊ/, /u/ in the carrier phrase "It's a /hVd/ again". The recordings were digitized at 10 kHz and F0 (fundamental frequency), F1 and F2 (vowel formants), SPL (sound pressure level), H1-H2 ("the amplitude difference between the first two harmonics in the acoustic spectrum" (2964), which is correlated with perceived 'breathiness'—an increase in H1-H2 is perceived as more breathy), and vowel nucleus duration were measured for each token. The authors found post-activation "compression of F1 range and movement toward normative mean values of F2 in the front vowels for some subjects" (2965), and all of their subjects demonstrated decreases in average vowel duration from pre-implant measures. Fundamental frequency decreased for subject FA (who had demonstrated an unusually high F0 pre-implant), but this was not found in the other subjects. SPL decreased for all subjects, and H1-H2 increased (correlated with the perception that speech was less 'pressed') for all but subject FB, who showed a decrease. Examining the longitudinal trends, Perkell et al. tentatively conclude that certain parameters, such as H1-H2 and SPL, change together. The authors also examine parameter relations in terms of underlying articulatory mechanisms. Their results from this preliminary study lead them to suggest that the model of speech production which best accounts for the cochlear implant subjects' data includes the interaction of postural adjustments, complex physical linkages, and phonemic adjustments.

Svirsky et al. (1992) records three of the same subjects as the Perkell et al. study at 26 weeks (subject MA), 73 weeks (subject FA), and 53 weeks (subject FB) post-activation. The subjects were recorded speaking five blocks of ten repetitions each of the vowels /i/, /ɪ/, /ɛ/, /ɑ/, /ʊ/, /u/) in the carrier phrase "It's a /hVd/ again." The block was recorded in conditions of 24-hour deprivation (off1); immediately after turning the processor on (on1); 15 minutes after turning the processor on (on2); turning the speech processor back off (off2); and 15 minutes after turning the speech processor off (off3). In order to measure speech breathing parameters, subjects FA and FB were recorded reading the Rainbow Passage (**A.4**) during the off1, on1, and off2 conditions, and subject MA read a different passage during the same conditions. Subjects' lung volume was measured using an inductive plethysmograph[2]. The recordings were digitized at both 10 kHz (for speech analysis) and 312.5 Hz (for lung volume measurements). Measurements for each token included F0, F1, F2, SPL, and H1-H2, as well as vowel duration. Mean airflow was also measured. The authors found that SPL increased when the speech processor was turned off, airflow and H1-H2 always changed in the same direction, and the most significant parameter changes occurred when there was a change in processor status. The authors find their results support a dual role for auditory feedback: both long term calibration, demonstrated by the greater changes when the processor was first turned on after being off for 24 hours, and short-term feedback, demonstrated by the rapid changes when the speech processor was turned either off or on in the later conditions.

These two studies provide evidence that experience with a cochlear implant affects vowel duration, vowel formants, the breathiness correlate, fundamental frequency, and sound pressure level. These speech parameters changed with increased experience, and also were affected by both long-term and short-term auditory deprivation.

2.4.3 Sibilant production studies

Matthies, Svirsky, Lane, and Perkell (1994) and Matthies, Svirsky, Perkell and Lane (1996) examine sibilant production. Matthies et al. (1994) find /s-ʃ/ production an interesting area for study as "the articulation required is relatively precise and complicated, there are few visual cues available, and the perception of all of the possible spectral information (up to 10 kHz) may not be possible within the frequency range of the cochlear implant" (1367). Five subjects (MB and MC, male, and FA, FB, and FC, female) were recorded twice before implantation and six times after implant activation: immediately at activation, and 4, 12, 26, 52, and 104 weeks post-activation. The subjects were recorded speaking three repetitions of the phrases "It's a /sad/ again" and "It's a /ʃad/ again" embedded in a larger corpus. The tokens were digitized with a 12 kHz sampling rate and low pass filtered at 5.5 kHz. The spectral shapes of the /s/ and /ʃ/ tokens were analyzed using FFT and normalized power spectra. The authors found contrast maintenance in subjects FA and MB, contrast improvement in subjects FB and MC over a 6 month period, and contrast improvement in subject FC (whose preimplant speech showed the poorest sibilant contrast) only after 6 months of implant use. The authors played digitized samples of pre- and postimplant /s/ from subject FB through an Ineraid implant and found an increase in the energy in electrodes 3 and 4 in the post-implant /s/, implying that subject FB could distinguish her pre- and postimplant /s/.

In Matthies et al. 1996, the authors used an EMMA[3] to measure the articulatory positions of a male cochlear implant subject during sibilant production, in conditions of both speech processor on and speech processor off (on1, speech processor was on from before the start of the experiment; off1, immediately after the speech processor was switched off; off2, the speech processor had been off for 15 minutes; on2, immediately after the speech processor was turned back on). The sibilant productions were digitized at 10 kHz and analyzed in the same manner as Matthies et al. 1994. Acoustic analysis shows that for this subject, the /s-ʃ/ contrast is sensitive to the speech processor being turned off–the /ʃ/ median slowly increases in frequency (drifts) in the off conditions, and changes sharply when the implant is turned back on. Articulatory data reveals an "anterior drift of the tongue-front transducer in the off2 condition" (944), consistent with their acoustic finding of /ʃ/ drift. Given the acoustic and articulatory changes between off2 and on2, the authors conclude that the subject was demonstrating the use of auditory feedback.

These two studies provide evidence that sibilant production is sensitive to experience with a cochlear implant, though one subject required a full six months of experience to see changes. The EMMA study in particular demonstrates sensitivity to short-term auditory deprivation, though those results should be considered in light of the single subject.

2.4.4 Stop production studies

Lane, Wozniak, and Perkell (1994) and Lane, Wozniak, Matthies, Svirsky, and Perkell (1995) examine voice onset time and syllable duration. Lane et al. 1994 recorded one male (MC) and three female subjects (FA, FB, FC) twice before speech processor activation, immediately at activation, and 4, 12, 26, 52, and 104 weeks postactivation. The subjects were recorded speaking three repetitions each of /b/, /d/, /g/, /p/, /t/, /k/ in the carrier sentence "It's a /Cad/ again." in a random sequence. The recordings were digitized at 10 kHz and VOT and syllable duration were calculated. For all subjects, syllable durations changed significantly with activation of the cochlear implant (generally becoming shorter); thus, to prevent confounding changes in speaking rate and changes in VOT, the authors normalized the later VOTs (indicated by VOTc) to the appropriate voiced or voiceless mean from the first baseline recording session. Three of the four subjects showed an increase in VOTc towards normative values following implant processor activation. The authors did find some evidence for place effect in the VOTcs of their subjects; an average of 6 msec separated adjacent places of articulation, and these differences were statistically significant.

In their first experiment, Lane et al. 1995 recorded the same subjects as Lane et al. 1994 (MC, FA, FB, FC), along with an additional male subject (MD), using the same materials and methods. In addition to measures of VOT and syllable duration, they also measured F0, H1-H2, and SPL. They again normalized the VOT for possible post-implant rate changes (VOTc). In their second experiment, subjects FA and FB disconnected their speech processors for 24 hours prior to recording, and were recorded speaking ten repetitions each of /t/ and /d/ in the "It's a /Cad/ again" carrier phrase, randomly mixed into a larger set of materials. As in Svirsky et al. 1992, this block was recorded in conditions of 24-hour deprivation (OFF1); immediately after turning the processor on (ON1); 15 minutes after turning the processor on (ON2); turning the speech processor back off (OFF2); and 15 minutes after turning the speech processor off (OFF3). The recording, digitizing, and measuring were the same as their first experiment. The on VOTs were normalized to the experiment 1 baseline off VOTs to correct any rate changes. The longitudinal results show changes in VOT patterned with changes in postural correlates (F0, SPL and H1-H2). The short-term deprivation experiment showed changes consistent with the longitudinal study. The authors suggest that "VOTc changes in voiced plosives with changes in hearing status may reflect predominantly postural changes while VOTc changes in voiceless plosives may reflect predominantly auditory validation of phonemic settings" (3105), as voiced VOTc changes correlated well with concurrent postural changes, while voiceless VOT changes were not observed with short-term deprivation.

These two studies provide evidence that experience with a cochlear implant has the greatest effect on syllable duration, with some effect being seen in increased normalized VOTs. Short-term changes in processor state primarily affected SPL, the breathiness correlate, and F0; these results are consistent with the vowel production studies.

2.4.5 Intonation study

Lane, Wozniak, Matthies, Svirsky, Perkell, O'Connell, and Manzella (1997) examine the role of auditory feedback on F0 and SPL. They recorded four subjects with cochlear implants (one male (CMD), three female (CFA, CFB, CFC)) and one female subject with an auditory brainstem implant (following damage from removing an auditory nerve tumor resulting from Neurofibromatosis-2) (NFA). Subjects were recorded twice before surgical intervention and at variable intervals afterwards reading the Rainbow Passage three or five times each. Lung volume was measured using an inductive plethysmograph. For this study, subjects CFA and CFB's 1 and 2 year recordings, subjects CFC and CMD's 6 month and 1 year recordings, and subject NFA's 11 week and 35 week recordings were analyzed. Each recording was divided into breath groups (using the plethysmograph data) and matched pre- and postintervention, and the SPL and F0 of each vowel was calculated. The contours of each breath group were normalized and their regularity was calculated. Comparing pre- and postintervention results, three of the four speakers with cochlear implants showed less variability in their F0 and SPL contours (the one who did not show less variability had the least variable contours preimplant) and subject NFA, who experienced a large reduction in auditory feedback, showed complementary results of greater variability in those contours. The authors find these results, along with a literature review of studies of SPL, F0, duration, F0 inflection, and intelligibility, support their theory that one role of auditory feedback is to "monitor transmission conditions, leading the speaker to make changes in speech postures aimed at maintaining intelligibility" (2244).

This study provides further evidence that F0 and SPL are sensitive to experience with a cochlear implant, which is consistent with the results of the vowel production and VOT studies.

2.4.6 Summary of the Perkell group acoustic production studies

Examining these studies as a whole, the authors have found evidence that the Ineraid four-channel cochlear implant provides a sufficient degree of auditory feedback to affect vowel placement and duration, /s-ʃ/ sibilant contrast, VOT and syllable duration, and what they term postural correlates, particularly F0 and SPL. The auditory feedback from the implant is seen to have short-term as well as long-term effects, as evidenced by the 24-hour deprivation off/on/off studies.

2.4.7 Recent Perkell group studies

In 2001, Perkell's research group published two articles examining vowel production and perception in users of the Clarion cochlear implant (see **2.2.2.3** for details of this implant's electrode array and speech processing strategies).

Lane, Matthies, Perkell, Vick, and Zandipour (2001) examine the role of hearing status on vowel space and coarticulation. They recorded seven postlingually deaf adults, three males (ME, MF, MG) and four females (FD, FE, FF, FG) (as well as one male and one female with normal hearing), three times before implant activation and at five intervals post-activation (approximately 0, 4, 12, 24, and 48 weeks). The subjects were recorded speaking ten repetitions of the vowels (/i/, /ɪ/, /ɛ/, /æ/, /ɑ/, /ʌ/, /ɔ/, /u/) in the frame sentences "Say /bVt/ again" and "Say /dVt/ again." The tokens were digitized at 10 kHz, low pass filtered at 4.5 kHz, and displayed using MITSYN. Vowel duration and midpoint were measured, and F1 and F2 (determined from LPC analysis) were recorded at 25 msec intervals between vowel onset and offset. Index of coarticulation (ratio of F2 at vowel onset to F2 at vowel midpoint) and average Euclidean vowel space were calculated for the last two recording sessions preceding implant activation and the last two sessions after implant activation. The researchers found little change in coarticulation with cochlear implant experience, and that the cochlear implant users maintained approximately the same vowel space as the hearing control speakers.

Vick, Lane, Perkell, Matthies, Gould, and Zandipour (2001) investigate the relationships between vowel perception and production in cochlear implant users, as well as the perception of CI users speech by hearing listeners. They recorded the same speakers as Lane et al 2001 (with the addition of one postlingually deaf male, MI) three or four times before implant activation and at five intervals post-activation (1, 4, 12, 24, 52 weeks), speaking ten repetitions of the vowels (/i/, /ɪ/, /ɛ/, /æ/, /ɑ/, /ʌ/, /ɔ/, /u/) in the frame sentence "Say /bVt/ again." The tokens were digitized at 10 kHz, low pass filtered at 4.5 kHz, and displayed using MITSYN. F1 and F2 at vowel midpoint were measured using LPC analysis for the data from the last two sessions before implant activation (Pre) and the 24 and 52 week sessions (Post). The researchers used linear discriminant analysis to generate a vowel confusion matrix for the pre- and post-implant sessions (note that due to cross-subject similarities in pronunciation, /ɔ/ was dropped from analysis). The eight cochlear implant users also participated in a perceptual experiment before each recording session. A normal hearing male and female were recorded speaking the eight vowels in /bVt/ context, and the cochlear implant subjects listened to two to three gender-matched 48 item blocks (8 vowels X 3 repetitions X 2 presentations) in each session, indicating their answers using a computer interface. In a third experiment, seventeen normally hearing adults (5 male, 12 female) listened to the CI users and hearing controls speech, such that a panel of ten participants listened to each speaker. Blocks consisting of 192 speech-shaped noise masked tokens (8 vowels X 3 presentations X 2 repetitions X 4 recordings) were created from each CI speaker's recordings and presented to the normal hearing adults via a computer interface. Comparing the results of the perception, production, and vowel recognition experiments, the researchers found that "improvements in the 8 implant users' perception of vowel contrasts...were accompanied in 4 speakers by increases in produced contrast distances... and in 6 speakers by increases in masked word recognition scores" (1263). For three speakers, their vowel productions were sufficiently distinct preimplant that recognition scores did not improve despite their improvements in vowel perception. Vick et al. find that cochlear implants provide their subjects with sufficient self-monitoring to adjust vowel production for greater contrast and intelligibility.

2.5 OTHER ACOUSTIC STUDIES

There are two other primary sets of acoustic studies of the speech of postlingually deafened adults with cochlear implants, each of which focus on speakers who received the Nucleus-22 cochlear implant (See **2.2.2.2** for descriptions of the specific speech processors and speech processing strategies) and speak languages other than U.S. English. The Langereis group papers, published between 1995 and 1998, study vowels, nasality, and fundamental frequency in native speakers of Dutch. Kishon-Rabin, Taitelbaum, Tobin, and Hildesheimer (1999) study vowels, VOT, fundamental frequency, sibilants, and word and sentence duration in native speakers of Modern Hebrew.

2.5.1 The Langereis group studies

The Langereis group studies of vowels (Langereis, Bosman, van Olphen, and Smoorenburg 1995, 1997) and fundamental frequency (Langereis, Bosman, van Olphen, and Smoorenburg 1998) are motivated by the limited number of studies (with few subjects) that have been published on vowel formants and fundamental frequency in cochlear implant users. Langereis et al. 1995, 1997, and 1998 use data from a significantly larger subject population (twenty subjects, twelve female and eight male) than previous studies. All of their subjects are postlingually deafened native speakers of Dutch who received the Nucleus-22 cochlear implant, with 3 using WSP and 17 using MSP processors. At twelve months post-activation, fifteen subjects were using the MPEAK stimulus strategy, four were using $F_0/F_1/F_2$, and one was using F_0/F_2. Speech recordings were made prior to implantation and 3 and 12 months post-implantation with the implant both on and off. Vowel materials consisted of the eleven Dutch monophthongs in hVt context; fundamental frequency materials consisted of an approximately 40 second text. Recordings were digitized at 10 kHz and analyzed using Entropic software. For vowels, F1 and F2 were measured using LPC analysis; for fundamental frequency, F0 was estimated using LPC analysis. Their results generally match well with those of the Perkell group; their subjects show a great deal of individual variation, with tendencies towards normative values over time. Subjects with an unusually high F0 or small formant range tended to show the most normalization.

Their study of nasality (Langereis, Dejonckere, van Olphen and Smoorenberg 1995) is motivated by previous studies which found evidence of deviant nasality in the postlingually deaf (309). Their subjects consist of 21 postlingually deafened native speakers of Dutch, 6 males and 15 females, with all subjects using MSP processors (this group of subjects partially overlaps with that of the Langereis et al. studies of vowels and fundamental frequency). At twelve months post-activation, three subjects were using the $F_0/F_1/F_2$ strategy and the rest were using MPEAK. As in the other studies, recordings were made prior to implantation and 3 and 12 months post-implantation with the implant both on and off. Materials consisted of a "standard text containing both oral and nasal phonemes" (310) as well as two sentences without nasal phonemes. Nasality was measured using the Kay Electronics model 6200 Nasometer and computed for each subject and condition. The authors found that at the pre-implant session most of their subjects "showed no effect of postlingual deafness on the degree of nasality in speech," with one subject found to be hyponasal and two hypernasal (312). They found no significant effect of implant experience on nasality, although the three subjects with nasality outside the normative range evidenced some normalizing effects at three months post-implant.

2.5.2 Kishon-Rabin et al.

Kishon-Rabin et al. investigate the effect of auditory feedback on speech production with a particular focus on the relationship between acoustic analysis and speech quality ratings, longitudinal changes beyond one year of cochlear implant use, and changes in speech production for non-English speakers. They use data from five postlingually deafened adult native speakers of Modern Hebrew, who were implanted with the 22-channel Nucleus cochlear implant and used the $F_0/F_1/F_2$ processor. Materials consisted of 50 monosyllabic minimal-pair words, 12 "every-day sentences" (7 declarative, 3 question, 2 imperative) and spontaneous speech. Materials were recorded prior to implantation and one, six, and 24 months post-activation. Recordings were digitized at 20 kHz and analyzed using Sensimetric Speech Station Ver. 2.1. The authors measured F1 and F2 of vowels (using LPC analysis), F0 of word-in-isolation and word-in-sentence (using a built-in autocorrelation function), VOT (using waveform information), sibilant spectral range (using a 512 point Fourier spectrum display), and word and sentence duration (using waveform and spectrogram displays). Speech was also rated for general speech and voice quality by speech therapists inexperienced in deaf speech. Kishon-Rabin et al. found significance for decreases in F0, word and sentence duration, and F1; increase of voicing lead in voiced plosives (closer to Hebrew norms); and changes in fricative spectral range. At two years post-implant most measured values fell within Hebrew norms. The speech therapists tended to judge preimplant as "worst", 2 years post-activation at "best", and 6 months post-implant as better than one month post-implant. The authors find that their results support Perkell's 1992 model, and suggest that further work should be done to examine changes in F1 and VOT in both languages more similar to Hebrew and languages more similar to English.

2.5.3 Summary of other acoustic studies

This group of studies provides evidence that fundamental frequency, vowel formants, VOT, and sibilant spectral range are sensitive to experience with a Nucleus cochlear implant, for subjects who speak languages other than English. These results match well with those of the Perkell studies.

2.6 OVERVIEW OF COCHLEAR IMPLANT STUDIES

Looking at the Perkell, Langereis, and Kishon-Rabin studies as a whole, there is clear evidence that the feedback provided by cochlear implants can affect vowel formants and duration, /s-ʃ/ sibilant contrast and spectral range, VOT and syllable duration, and fundamental frequency. The subjects in most these studies, however, did not have access to the current range of digital cochlear implant technology. Current processors transmit a far greater frequency range and far more spectral detail than the Ineraid processor used by the subjects in the 1990s MIT studies. Looking at the Nucleus processors used by subjects in the Langereis and Kishon-Rabin studies, the MPEAK and $F_0/F_1/F_2$ processing strategies extract a small set of linguistically relevant features from the acoustic signal, while the more recent SPEAK and ACE strategies operate in a very different fashion, presenting an overall picture of the acoustic signal.

2.7 THE ROLE OF FEEDBACK IN SPEECH PRODUCTION

2.7.1 Lane and Tranel

No one questions the developmental role of auditory feedback; what, however, is the role of auditory feedback in the speech production of adults?

This question has been widely debated since the early 1970s, when Lane and Tranel (1971) proposed that auditory feedback serves a social function, rather than a regulatory one. Starting with Lombard's 1911 observation that "a speaker changes his voice level similarly when the ambient noise level increases, on the one hand, and when the level at which he hears his own voice (his sidetone) decreases, on the other" (677), Lane and Tranel review the resulting sixty years of research into this phenomenon and argue against the widely-held conclusion that the Lombard sign is "evidence that the speaking mechanism is a servomechanism controlled by auditory feedback" (692). The authors' review of the literature, along with their own research, leads them to conclude that auditory feedback is used "largely to maintain effective communication under varying external conditions" (693).

The following papers, published between 1974 and 1998, expand on Lane and Tranel's observations. Siegel and Pick (1974) manipulate the intensity of auditory feedback, both with and without added noise, and find that their evidence supports Lane and Tranel: auditory feedback "does not begin to regulate speech except under circumstances in which communication is difficult" (1624). McClean (1977) measures lip movement in conditions of unpredictably presented masking noise, and finds that masking caused significant changes in lip velocity. He concludes that his data is not entirely predicted by Lane and Tranel's explanation: for two subjects, masking caused decreases in displacement (and thus decreases in intensity) rather than the predicted increases in intensity. Van Summers, Pisoni, Bernacki, Pedlow, and Stokes (1988) find that speech produced in noise demonstrates segmental as well as prosodic differences from speech produced in quiet; formant frequencies and short-term spectra of vowels were most affected. Their perceptual experiments demonstrate that the speech produced in noise was more intelligible than speech produced in quiet, supporting Lane and Tranel.

Garber and Moller (1979) observe that subjects decrease nasalization when hearing their voices low-pass filtered as they speak, perhaps compensating for increased perceived nasality; they find that Lane and Tranel's explanation does not work as well in this case "since it is unclear that decreasing nasalization would increase intelligibility, especially for normal speakers" (329). They conclude that nasalization is under feedback control. Elman (1981) presents subjects with frequency-shifted feedback, and finds that subjects adjust their F0 up or down to compensate. In recent work on pitch-shifted feedback, Burnett, Freedland, Larson, and Hain (1998) find that their results, along with the results from Elman's study and other similar studies, support "the hypothesis that auditory feedback is important for voice F0 control" (3160), though their results are complex: some subjects shift their fundamental frequency in the same direction as the pitch-shift stimulus, and some shift in the opposite direction. These studies all demonstrate production changes resulting from the manipulation of perceptual feedback; the question is whether these changes are all functions of maintaining communication, or are some of them under more direct auditory feedback control? There is also concern whether these sorts of experiments can be compared to real-world situations.

2.7.2 Intelligibility and acoustic studies of deaf and hearing impaired speech

Studies of deaf and hearing impaired speech can provide this real-world evidence, as these speakers have all lost some or all auditory feedback due to their hearing impairments. In an intelligibility study on the speech of postlingually deafened adults, Cowie, Douglas-Cowie, and Kerr (1982) find varying degrees of 'speech deterioration' among their subjects, with some subjects showing little deterioration and others showing impaired intelligibility. Goehl and Kaufman (1984), on the other hand, had speech-language clinicians rate the speech of five adventitiously (postlingually) deafened adults and five age-matched hearing controls. They found that the judges, when asked if articulation was within normal limits and whether the speaker could be identified as deafened, did not distinguish the two groups. Further analysis revealed that the deaf subjects were more likely to be identified as deaf, and these judgments may have resulted from factors such as 'rate' and 'voice quality' (63). It is clear that the degree to which speech intelligibility is reduced varies greatly.

Turning to acoustic studies of deaf and hearing impaired speech, Lane and Webster (1991) examine the speech of postlingually deafened adults. They found that their subjects produce greater variations in pitch than comparable hearing subjects and show a lesser degree of place differentiation in fricative and stop consonants. Waldstein (1990) recorded six postlingually deafened adults and one postlingually deafened teenager (and seven age/gender-matched hearing controls) reading tokens designed to produce a variety of stop consonants, vowels, and intonation patterns. Acoustic measurements revealed that her postlingually deafened subjects have a more restricted vowel space and tend to produce vowels that are longer in duration (and have more variability in duration) than comparable hearing subjects, and for some subjects, F0 contrasts are exaggerated. Rubin-Spitz and McGarr (1990) recorded eight deaf children (and one female hearing adult) reading a sentence list consisting of declarative sentences, wh-questions, and yes-no questions. They found "little if any differentiation between mean F0 of declarative and nondeclarative utterances by the deaf talkers" (178) and their perceptual experiment indicated that the perception of terminal falls in the speech of deaf children seems to be correlated with the rate of frequency change (e.g., the dynamic nature of the contour), rather than the absolute change in frequency.

2.7.3 Perkell's two-level model of speech production

The Perkell group studies of the speech of cochlear implant users, along with these articles, contribute to a two-level model of speech production. One level is unconscious but active, and corresponds to what the Perkell group calls "postural correlates." This level includes articulatory motor habits, as well as "the balance between expiratory and inspiratory forces associated with a subglottal pressure, average tension of the vocal folds, average degree of adduction of the glottis, average position of the tongue body, and speaking rate" (Lane et al. 1995, 3096). This unconscious but active level responds to auditory feedback as described by Lane and Tranel and subsequent studies; the speech postures change to maintain effective communication. The other level involves segmental changes as conscious or semi-conscious responses to auditory feedback. Garber and Moller's results with nasals would fall into the segmental level, as would the findings on sibilants by Matthies et al. (1994, 1996). The studies of deaf speech seem to provide evidence that the loss of auditory feedback primarily affects the postural level (changes in vowel space, vowel duration, and intonation contours, changes in timing). This loss, however, may also effect the segmental level (loss of place differentiation in stops and fricative consonants). The Perkell group finds their acoustic studies of postlingually deafened adults with four-channel Ineraid implants primarily provide evidence for postural effects of auditory feedback (vowel placement and duration, syllable duration, fundamental frequency, VOT), though the sibilant studies provide evidence for segmental effects.

2.8 A NEW STUDY OF THE SPEECH OF ADULTS WITH MODERN COCHLEAR IMPLANTS

Changes in processor and implant technology have long-ranging implications in terms of possible improvements in speech production. The greater frequency range and increased spectral detail should particularly have an effect on vowel placement (as more specific detail about the formants should come through in the spectral peaks), sibilants (as the majority of sibilant noise is transmitted in the area of the spectrum above the 3400 Hz maximum of the Ineraid implant), and intonation (given the greater range of spectral detail also transmitted for lower frequencies). Given these implications, an acoustic study of the speech of postlingually deafened adults with current cochlear implants was needed.

NOTES

1 Newspaper article reviewing cochlear implant studies.

2 An inductive plethysmograph is used to measure lung volume. It consists of two wired cloth bands, one placed around the ribcage and the other at the waist. The bands are connected to an oscillator module which feeds into a signal demodulator that "outputs signals proportional to the cross section of each coil...it is assumed that the cross-sectional area within the rib cage and the abdomen coil, respectively, reflects all of the changes occurring within the respective lung compartment, and further that the lung volume change is the sum of the volume changes of the two compartments. Under optimal situations lung volume can be approximated with an error less than 10%" (Asker 2001).

3 An EMMA, an electromagnetic midsagittal articulometer, "uses alternating magnetic fields that induce voltages in small transducer coils mounted on the subject's articulators" (Matthies et al. 1996, 938). These movement signals are recorded and used to determine articulatory changes at different points in the subject's speech productions.

3 Methodology

3.1 SUBJECTS

Subject AM is a 34-year-old man who lives in Northern California. His hearing loss probably started around age 1, from either an infection or the ototoxic antibiotics used to alleviate the infection. His hearing loss was diagnosed and he first received a hearing aid in his right ear in kindergarten, with a hearing aid for the left ear following in second grade. He acquired speech normally, and was entirely mainstreamed in school. His hearing has been gradually degrading; as of age 29, he could still do telephone-based technical support, but currently his hearing loss is classified as severe to profound in the left ear (80 dB) and profound in the right ear (110 dB). He received a Clarion CII cochlear implant in his right ear on January 9, 2002 and it was activated February 5-6, 2002. He primarily uses the CII BTE processor. AM was recorded on 1/3/02, 3/9/02, 5/17/02, and 8/18/02. AM describes the implant activation process as like getting glasses—suddenly getting something back, rather than something totally new. When his implant was first activated, speech had overtones of hoarseness or decay effects, depending on what processing strategy he used. He still finds speech somewhat distorted at six months post-activation. Generally AM is pleased with his implant, though he finds music much more difficult to listen to, and relies on lipreading more, than when he was using hearing aids. I would describe AM's speech pre-implant as clear and understandable, with more signs of vocal effort and changes in volume compared to a normal hearing speaker. At six months post-activation, AM's speech has not changed significantly from what it sounded like pre-implant.

Subject DS is a 53-year-old woman who lives in the Midwest. Her binaural progressive hearing loss started approximately ten years before implant surgery, and has been diagnosed as having a genetic cause. Her hearing loss was mild to begin with and became more severe over time. Her hearing loss is classified as profound in the left ear and severe in the right. She has found hearing aids unsatisfactory and frustrating. She received a Nucleus N-24 implant in her left ear on April 2, 2002 and it was activated May 7-8, 2002. She primarily uses the ESPrit ear-level processor. DS was recorded on 3/21/02, 6/13/02, 8/6/02, and 11/5/02. When DS's implant was first activated, people sounded like "little munchkins," but she was able to understand her husband's voice right away. By six months, she describes the sound from the implant as free of distortion. DS is extremely pleased with her cochlear implant. I would describe DS's speech pre-implant as slow and deliberate, with a nasal overtone. At six months post-activation, DS's speech rate has increased dramatically, and her speech is less nasal.

Subject MS is a 33-year-old woman who lives in the Midwest. Her progressive binaural hearing loss of unknown etiology was initially diagnosed at age 21, and in the four years prior to implantation, her hearing loss had progressed to severe loss on the left and profound on the right. She received a Nucleus N-24 cochlear implant in her left ear on December 30, 1999 and it was activated January 24-25, 2000. She primarily uses the ESPrit ear-level processor. MS was recorded on 12/18/99, 2/26/00, 4/29/00, and 7/29/00. When her cochlear implant was first activated, MS told the audiologist to turn it off because not only did everything sound high-pitched and squeaky, there was a time delay between when somebody would speak and when she actually heard what they were saying, "like a badly dubbed movie." By three months, MS describes things as sounding the way they did before she lost her hearing, though she does comment that it is difficult to hear the pitch of her own voice. MS is very pleased with her cochlear implant experience, and finds it has had a profound effect on her life. I would describe MS's speech pre-implant as nasal in quality, and somewhat monotonous. At six months post-activation, MS's speech is less nasal and shows more variation in pitch.

3.2 ACOUSTIC ANALYSIS

3.2.1 Materials and procedures

Stimuli for vowel analysis consisted of the vowels (/i/, /ɪ/, /ɛ/, /æ/, /ɑ/, /ʌ/, /ɔ/, /ʊ/, /u/) in the carrier phrase "It's a /hVd/ again." (see **A.1**). The stimuli were presented on index cards. Three repetitions of each word, in random order, were elicited. (following Lane et al. 1992).

Stimuli for stop and future fricative analysis consisted of CaC(C) words (as in Waldstein 1990) with the initial consonants [p t k b d g] (five words beginning with each of these consonants, with the final consonant varied in order to create real words, see **A.2**) and /sad/, /ʃad/ (following Matthies et al. 1994), /fat/, /vat/ in the carrier phrase "It's a _____ again." The stimuli were presented on index cards. Three repetitions of each word, in random order, were elicited.

For both the vowel and consonant stimuli, the frame sentence "It's a _____ again." was presented on a sheet of paper visible to the subject while the individual vowel and consonant stimuli were presented on index cards.

Stimuli for intonation analysis consisted of the phrase "Bev loves Bob", as a statement, statement with Bev, loves, or Bob emphasized, a question, and a question with Bev, loves, or Bob emphasized (following Atkinson 1973, as cited in Lieberman & Blumstein 1988, see **A.3**). Emphasis was indicated by underlining, statement was indicated by a period at the end of the phrase, and question was indicated by a question mark at the end of the phrase. The stimuli were presented on index cards. Three repetitions of each phrase, in random order, were elicited.

The Rainbow Passage (Fairbanks 1960, see **A.4**) was recorded for the purposes of intonation analysis. The passage was presented on a piece of paper located in front of the subject. Three repetitions of the passage were elicited, interspersed between the other sections of the recording session.

Before the interview section, subjects were asked to briefly tell a story depicted in a wordless children's book (Wiesner 1991), for future analysis. This was recorded once per session.

Subjects were recorded prior to their implant, and one month, three months, and six months after implant activation. For specific listings of the card presentation orders, see **Appendix B**. Subjects were asked interview questions at the end of each of these sessions and their answers were recorded (see **B.5, B.6, B.7**). The total recording time per session was approximately 20 minutes. Subjects were recorded in the Recording Studio at the Language Laboratories and Archives at the University of Chicago, a recording studio local to the subject, or a quiet room, using a digital recorder and a unidirectional microphone. Recording levels were adjusted for maximum signal without clipping.

3.2.2 Digitizing

Utterances were digitized at 20 kHz (MS) or 48 kHz (AM, DS) using Kay Elemetrics Computerized Speech Lab (CSL) and analyzed using CSL, WaveSurfer, and Praat.

3.2.3 Measurements used: Theoretical background

As neither intonation nor vowels provide physical landmarks to aid articulation, and stops have physical landmarks but require precise articulation and timing, these factors should be most affected by the loss of acoustic feedback as a result of hearing loss, and by subsequent recovery of (partial) hearing after implantation of a cochlear implant. Thus, acoustic analysis of these aspects of speech should provide the most information as to the articulatory changes that follow experience with a cochlear implant.

3.2.3.1 Vowel formants

Waldstein (1990)'s study of the speech of postlingually deaf adults found that "in general, the vowel spaces for these speakers are more restricted in range, and the vowels are less discretely positioned within the space, relative to normal" (2104). Perkell et al. (1992) found normative changes in vowel space for some of their subjects post-implant. Lane et al. (2001), on the other hand, found that the cochlear implant users in their study maintained approximately the same vowel space as hearing control speakers. Given the lack of physical landmarks to aid articulation, characterizing vowel space by measuring vowel formants should provide information as to articulatory changes related to auditory feedback from the cochlear implant.

3.2.3.2 Duration and VOT

Waldstein (1990) found that for vowels, "overall average duration values for the postlingually deafened subjects tend to be longer when compared to overall values for the hearing subjects" (2107); Perkell et al. (1992) found that for all of their subjects, average vowel duration decreased post-activation.

In terms of CV syllable duration and VOT, Waldstein (1990) finds her postlingually deafened subjects produce overall shorter VOT values relative to normal hearing speakers. Lane et al. (1994) found that changes in VOT varied with changes in CV syllable duration post-activation, requiring a correction factor to facilitate accurate comparison of pre-implant and post-activation VOTs. Lane et al. note that several features of VOT make it a good measure of articulatory changes related to auditory feedback:

> First, the deaf speaker cannot see or hear the laryngeal gestures that control VOT....Thus if VOT is normally regulated by hearing, it may be particularly vulnerable to deafening; and if it can be discriminated well with the aid of a cochlear implant, it may change following activation of the implant processor. Second, the control of VOT requires precise timing of glottal and superglottal events and the temporal coordination of articulatory events may be particularly vulnerable to deafening (56).

These factors all contributed to my decision to measure vowel duration, vowel word duration, VOT, and CV syllable duration.

3.2.3.3 Fundamental frequency

In studies using normal hearing speakers, Elman (1981) and Burnett et al. (1998) both found that subjects adjust their F0 up or down to compensate for F0-modified feedback. Waldstein (1990) found no overall effect of postlingual deafness on mean F0, but when looking at F0 maxima, did note that "some postlingually deafened subjects were occasionally less precise in controlling F0 excursions in the production of yes-no questions" (2110). Lane et al. (1997), in their fundamental frequency study which used the Rainbow Passage, found that most of their postlingually deaf speakers produced less variable average F0 contours post-activation. These factors contributed to my decision to measure peak F0 (e.g., F0 maxima) for the vowel tokens and Bev loves Bob sentences, and average F0 for the Rainbow Passage readings.

3.2.4 Measurement criteria

3.2.4.1 Vowels

Vowels were measured from wide band spectrograms. Vowel duration is defined as Start of V to End of V. Vowel word duration is defined as Start of H to End of word (see Figure 3.1–3.3).

Table 3.1 Vowel measurement criteria

Frequency	F1, F2, F3 determined visually	Measure center of formant, visually integrating, using information from formants in H if appropriate
Peak F0	Using automatic pitch extraction	Measure highest value within vowel
Duration	Start of H: ranked	1. start of high pitch fricative energy 2. end F2 of "a"
	Start of V	Onset of F2 of vowel
	End of V	End of F2 of vowel
	End of word: ranked	1. D burst 2. end of voicing of D 3. start of vowel of "Again"

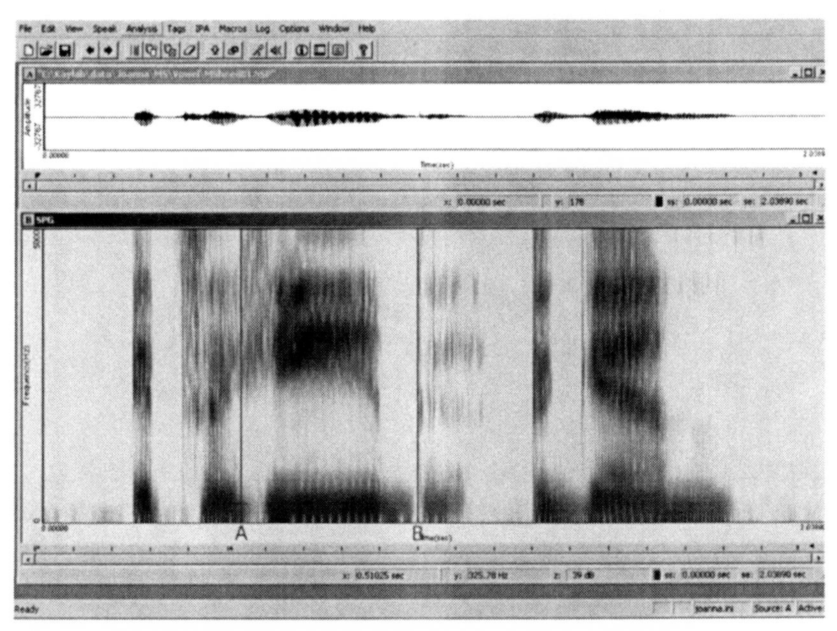

Figure 3.1 Subject MS, "It's a heed again," one month post-activation. Line A represents Vowel start of H, rank 1: Start of high pitched frication noise. Line B represents Vowel end of word, rank 1: D burst.

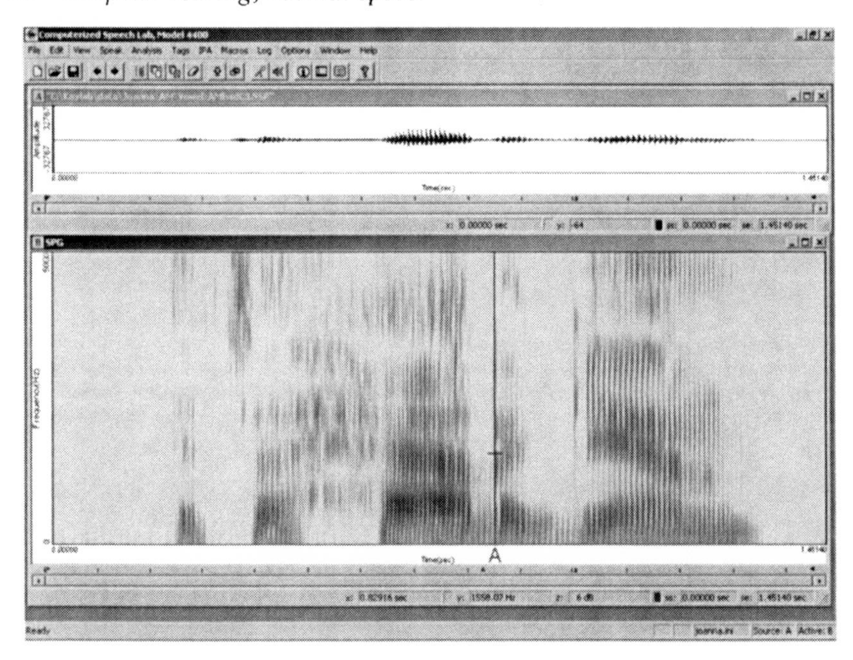

Figure 3.2 Subject AM, "It's a had again," three months post-activation. Line A represents vowel end of word, rank 3: start of vowel of "Again."

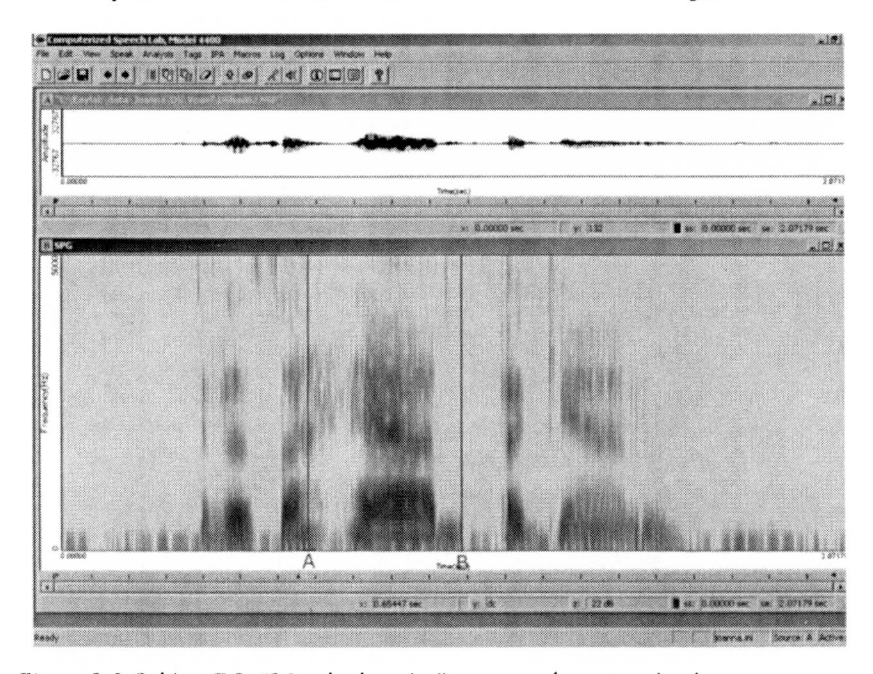

Figure 3.3 Subject DS, "It's a had again," one month post-activation. Line A represents vowel start of H, rank 2: end F2 of a. Line B represents vowel end of word, rank 3: End of voicing of D.

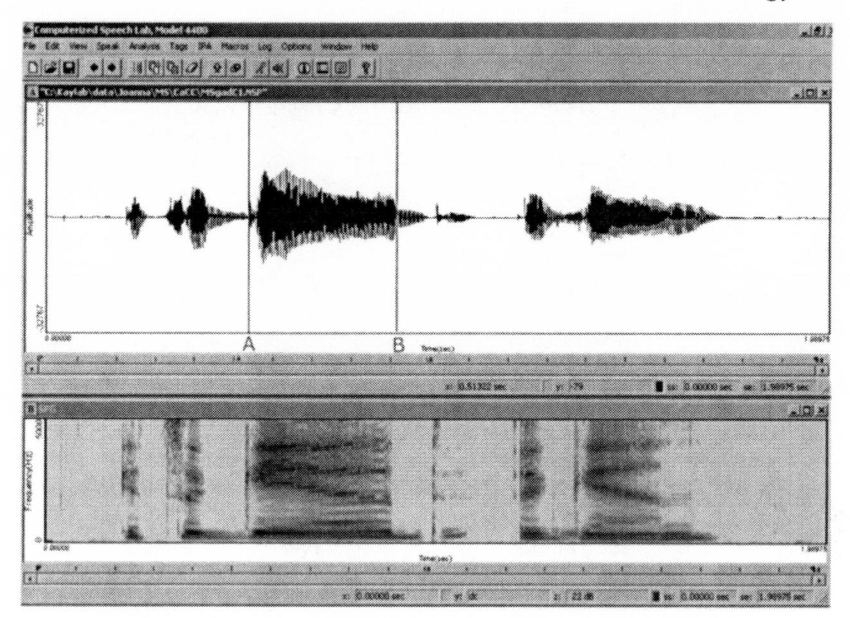

Figure 3.4 Subject MS, "It's a gad again," three months post-activation.
Line A represents the voiced through closure VOT=0 criteria; Line B represents end of vowel.

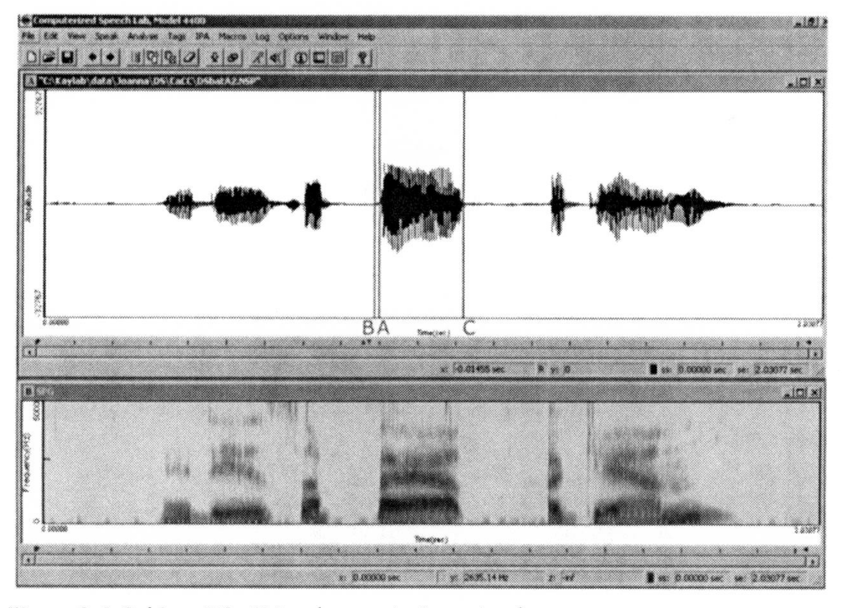

Figure 3.5 Subject DS, "It's a bat again," pre-implant.
Line A represents the burst release; line B represents the onset of voicing; line C represents end of vowel.

3.2.4.2 VOT

Stops were measured from the expanded waveform, and these measurements were checked against the spectrogram. VOT is defined as the interval between burst release and voicing onset. CV syllable duration is defined as the interval between burst release and end of vowel (see Figures 3.4–3.6).

Table 3.2 Stop measurement criteria

Burst release	Energy spike in waveform
Voicing onset	Start of regular voicing periods in waveform (start of first complete cycle). When voiced through closure, burst release and voicing onset measured at the same point (VOT=0)
End of vowel	End of regular voicing periods in waveform

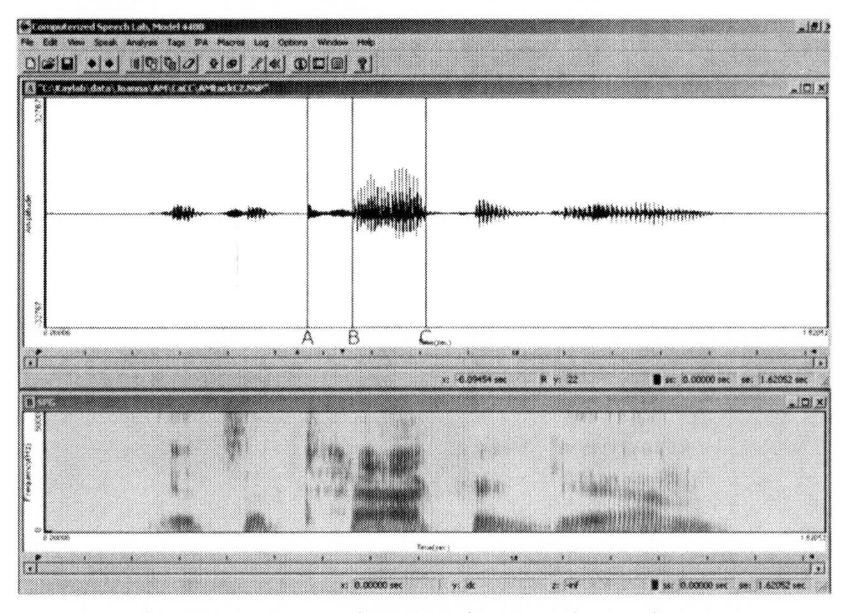

Figure 3.6 Subject AM, "It's a tack again," three month recording session. Line A represents burst release; line B represents onset of voicing; line C represents end of vowel.

3.2.4.3 Fundamental frequency

The Rainbow Passage was divided into the following phrases, by breath group:

1 When the sunlight strikes raindrops in the air,
2. they act like a prism and form a rainbow.
3. The rainbow is a division of white light into many beautiful colors.
4. These take the shape of a long round arch,
5 with its path high above,

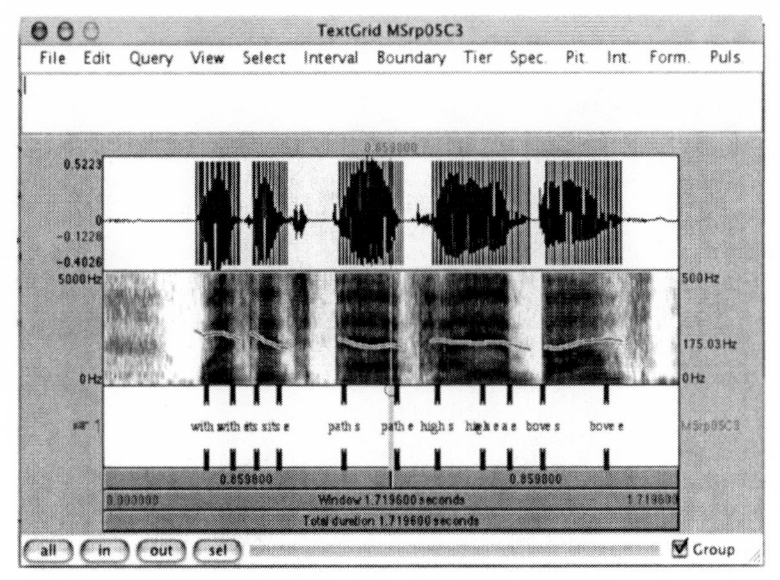

Figure 3.7 Subject MS, Rainbow Passage phrase 5, three months post-activation. Syllable boundaries are indicated.

6. and its two ends apparently beyond the horizon.
7. There is, according to legend,
8. a boiling pot of gold at one end.
9. People look, but no one ever finds it.
10. When a man looks for something beyond his reach,
11. his friends say he is looking for the pot of gold at the end of the rainbow.

Each phrase file was opened in Pratt, using a script (written by Fang Liu and Yi Xu) that generated a TextGrid window (**Figure 3.7**) linked to a window displaying the vocal pulses. Vocal pulses were examined and corrected if necessary, and the boundaries of each syllable were marked in the TextGrid window. Average F0 for each syllable was calculated using the resulting files, omitting areas of glottalization.

Fundamental frequency was measured using automatic pitch extraction, with the pitch marks displayed overlaying the spectrogram. For the Bev loves Bob sentences, the peak F0 of Bev, peak F0 of loves, and peak F0 of Bob were measured.

3.2.5 Data analysis

Subjects' data were compared to their own, and were analyzed for longitudinal trends as well as implant effects (pre- vs. post-implant). Individual subject trends were compared to those found in the MIT group papers. Subject interview responses were used to classify their data.

3.3 PERCEPTUAL STUDY

3.3.1 Materials and procedures

Stimuli for consonant recognition consist of the consonants /p t k b d g s ʃ/ in the frame sentence "It's a /Cad/ again". Stimuli for vowel recognition consist of the vowels [i ɪ ɛ æ ʌ ɑ ʊ u] in the frame sentence "It's a /hVd/ again."

Three repetitions of the consonant and vowel stimuli, spoken by a native male English speaker from Minnesota, were recorded and digitized at 20 kHz using CSL.

The consonant blocks consist of 48 randomized stimuli (2 repetitions X 3 presentations X 8 consonants). The vowel blocks consist of 48 randomized stimuli (2 repetitions X 3 presentations X 8 vowels).

Consonant and vowel blocks were presented via the ASPP perceptual test program (part of CSL) or via Inquisit. Prior to the first testing block, exemplars of both the consonants and vowels were presented. During each testing block, the computer played the stimuli (at a comfortable listening level from a small loudspeaker or a laptop internal speaker) and then the subject selected their response on the screen from the 8 possible /Cad/ words (for the consonant blocks) or the 8 possible /hVd/ words (for the vowel blocks). Three blocks each for consonant and vowel identification were presented per session. Perceptual testing sessions occurred in conjunction with the one month, three month, and six month post-activation recording sessions.

3.3.2 Analysis

Confusion matrices were constructed from the identification data for each of the consonant and vowel tests. This data was compared for each subject with the acoustic analysis of their session-matched stop and vowel production.

4 Vowels and vowel perception

4.1 PREVIOUS FINDINGS

Waldstein (1990) found that postlingually deafened subjects in general had a more restricted vowel space than hearing subjects. Comparing pre- and post-activation data, Perkell et al. (1992) found F1 decreased for all subjects, F2 tended to normalize, vowel duration decreased for all subjects, and there were fewer F0 differences between vowels. See **2.4.2** for a detailed analysis of this study.

4.2 VOWEL SPACE

4.2.1 Methods used in analyzing vowel space

The F1-F2 vowel space graphs in this section follow Lane et al. 2001. Average values from Peterson and Barney (1952) are indicated by dotted lines and included for purposes of comparison.

Lane et al. 2001 define average (pairwise) vowel spacing (AVS) as "the Euclidian distances in the F1-F2 plane between all 28 pairs of the 8 vowels...computed and averaged" using the following equation:

$$AVS = \frac{2}{n(n-1)}\left\{\sum_{i=1}^{n-1}\sum_{j=i+1}^{n}\left[\left(F1_i - F1_j\right)^2 + \left(F2_i - F2_j\right)^2\right]^{1/2}\right\}$$

"where n=number of vowel pairs, $F1_i$ and $F1_j$ are the first formants of the *ith* and *jth* vowels and $F2_i$ and $F2_j$ are the second formants of the *ith* and *jth* vowels" (555). Comparing pre- and post-activation sessions, Lane et al. found that for some subjects AVS increased, and for others it decreased (557).

Another method of calculating vowel space follows Bradlow, Torretta, and Pisoni (1996). The triangle formed from the values of [i], [a], and [u] in the F1-F2 formant plane is graphed, and the "Euclidean area covered by the mean of each vowel category" is calculated (263). Bradlow et al. hypothesized that greater triangle area would correlate with higher intelligibility, but did not find a positive correlation across the speakers in their study (264). However, vowel triangles serve as a useful metric for comparing changes in vowel articulation with cochlear implant experience.

4.2.2 Vowel space analysis

Subject AM's front vowels are closer to normative values with continued cochlear implant experience, and the shift upwards in F2 for [ɛ] is significant when comparing pre-implant and six months, as is the downward shift in F1 of [æ]. There is little change in his back vowels, with only the downward shift of F2 of [u] statistically significant. AM shows little distinction between [a] and [ɔ] on the F1-F2 formant plane; [u] and [ʊ] are also very similar to each other. Compared to Peterson and Barney values, AM's vowel space is more centralized.

Calculating AVS for AM's nine vowels at each recording interval, a pattern of change and return emerges: there is an increase in AVS at one month and again at three months, but the value at six months is closer to that at

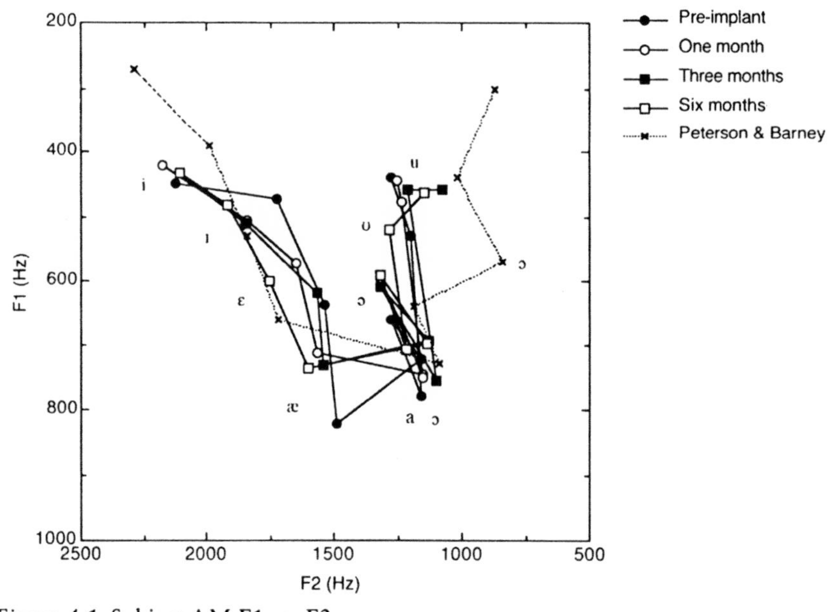

Figure 4.1 Subject AM F1 vs. F2

Table 4.1 AVS and triangle area, subject AM

AM	AVS (Hz)	Triangle area (Hz^2)
Pre-implant	434	120865
One month	464	138638
Three months	474	122164
Six months	466	113031

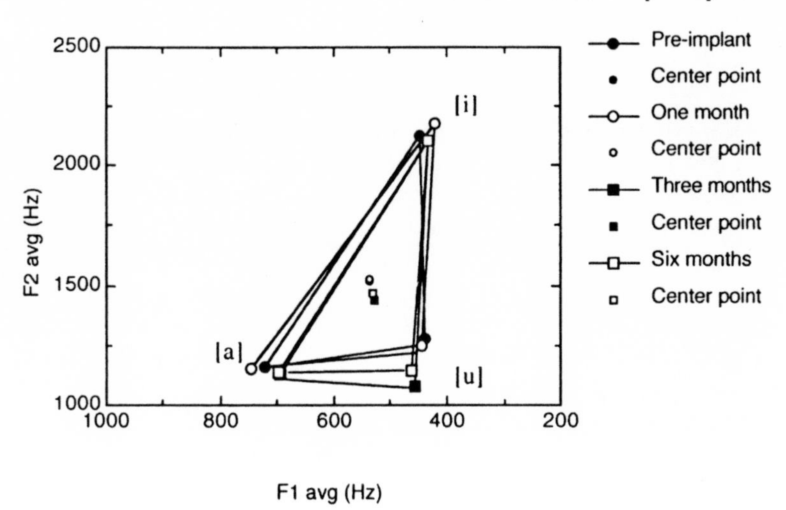

Figure 4.2 Subject AM [i]-[a]-[u] triangle, with center points of each triangle indicated

one month (**Table 4.1**). Changes in AVS are significant for the pre-implant vs. one month and pre-implant vs. three months comparisons.

The triangle area results follow a different pattern; increase at one month and decreases at three and six months. The changes in vowel triangle area for subject AM are not statistically significant. The shifts in F2 for the triangle center points are significant for pre-implant vs. three months, one month vs. three months, and one month vs. six months.

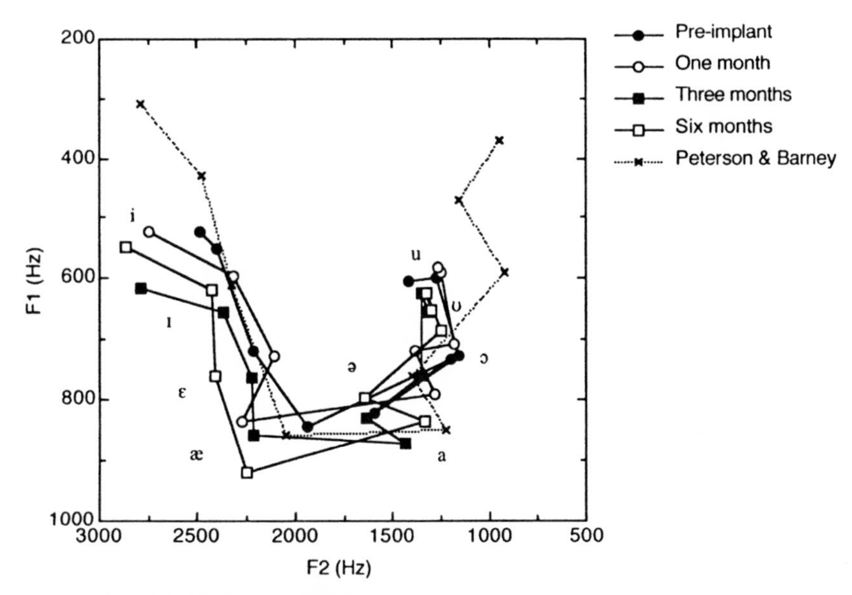

Figure 4.3 Subject DS F1 vs. F2

Subject DS shows an expansion of vowel space with cochlear implant experience, primarily in the front vowels. There is an upward shift in both F1 and F2 for front vowels. Increases in F1 and F2 are significant for [i] (F1 pre-implant vs. three months, one month vs. three months, three months vs. six months; F2 pre-implant vs. one month, pre-implant vs. six months, pre-implant vs. six months). The back vowels do not change significantly, with the exception of the central vowel [ə] which shows significant shifts in both F1 and F2 (F1 pre-implant vs. one month, pre-implant vs. six months, one month vs. six months; F2 pre-implant vs. one month, pre-implant vs. six months; one month vs. three months; one month vs. six months). At pre-implant, DS did not distinguish [a] and [ɔ]; post-implant, these two vowels are clearly distinguishable on the F1-F2 plane. [u] and [ʊ] appear very similar at all recording sessions. As with subject AM, DS's vowel space is centralized compared to Peterson & Barney values.

DS demonstrates a dramatic increase in vowel space post-implant, using both AVS (nine vowel) and triangle area measurement methods (Table 4.2). The increase in AVS is significant for pre-implant vs. six months and three months vs. six months.

The triangle area results appear to be much more dramatic, with the area at one month being over 2.5 times the pre-implant area:

Table 4.2 AVS and triangle area, subject DS

DS	*AVS (Hz)*	*Triangle area (Hz²)*
Pre-implant	667	59717
One month	728	157272
Three months	677	162851
Six months	763	146286

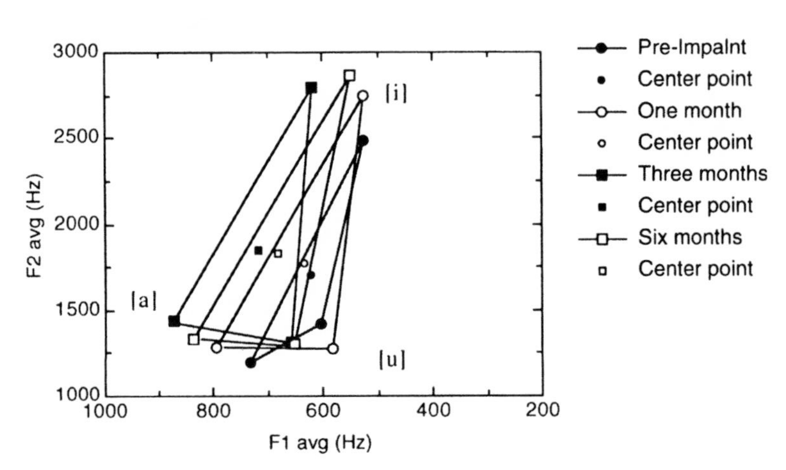

Figure 4.4 Subject DS [i]-[a]-[u] triangle, with center points of each triangle indicated

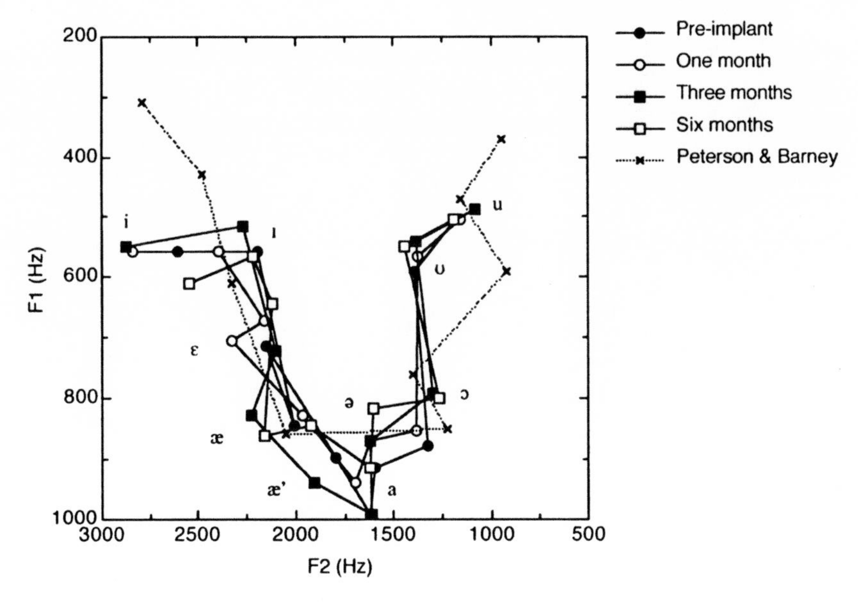

Figure 4.5 Subject MS F1 vs. F2

The changes in area, however, are only significant when comparing one month and three months (a comparison which is not significant when looking at AVS). The movement of the triangle center points is highly significant for F1 (all comparisons but pre-implant vs. one month) but not significant for F2. It is interesting that the vowel triangle area metric produces exaggerated results for this subject.

Unlike the other two subjects in this study, MS pronounces [æ] as a diphthong; in Figure 4.5, the first vowel of the diphthong is indicated by [æ] and the second by [æ']. [æ] is similar to [ɛ] in F2 but with a higher F1, while [æ'] is similar to [a] in F1 but with a higher F2. MS's back vowels are fairly stable, with only [a] showing a statistically significant shift (comparing pre-implant and one month for F2). The shifts in F2 of [i] are significant for all comparisons; for [ɪ] shifts in F2 are significant for pre-implant vs. one month, one month vs. three months, and one month vs. six months; for [ɛ] shifts in F2 are significant for pre-implant vs. one month, pre-implant vs. three months, and pre-implant vs. six months. MS makes a clear F1-F2 distinction between [a] and [ɔ] at all recording sessions. Of the three subjects in this study, MS has the greatest distinction between the back vowels [u] and [ʊ]. The overall shape of her vowel space is also most similar to the Peterson & Barney values, though her high vowels have a much higher F1.

For MS, AVS and the triangle area metric produce strikingly different results. (I note that her AVS was calculated using ten vowels, to account for the diphthongized [æ]).)

Table 4.3 AVS and triangle area, subject MS

MS	*AVS (Hz)*	*Triangle area (Hz^2)*
Pre-implant	607	334926
One month	699	352418
Three months	718	437421
Six months	604	253657

MS's triangle area is much greater than that of the other two subjects; her highest area (at three months) is more than double those of the other two. But looking at her AVS, it becomes clear that the high areas for the triangles are an artifact of that method, and do not reflect a larger overall vowel space.

Shifts in AVS are significant for pre-implant vs. one month, pre-implant vs. three months, one month vs. six months, and three months vs. six months. Shifts in triangle area are significant for pre-implant vs. three months, pre-implant vs. six months, and three months vs. six months. There is more crossover of significance with these measures for MS than for the other two subjects. The movement of the triangle center points is significant for F2 for pre-implant vs. one month and one month vs. six months. MS's AVS results point to a pattern of change and return: AVS increases at one and three months, but at six months it is nearly identical to the pre-implant value.

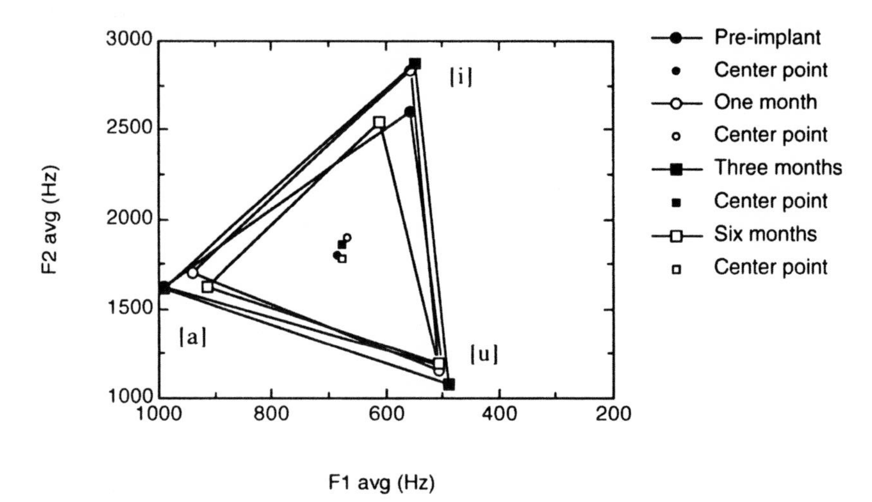

Figure 4.6 Subject MS [i]-[a]-[u] triangle, with center points of each triangle indicated

4.2.3 Comparison with Perkell et al. 1992

Unlike the findings in Perkell et al., there was no change in F1 that occurred for all subjects. In addition, F2 tended to change upward in frequency, and this change is in a normative direction for AM and DS.

4.3 VOWEL AND VOWEL WORD DURATION

4.3.1 Analysis of vowel and vowel word duration

For the short vowels [ɪ], [ʊ], [ə], and [ɛ], AM's vowel duration is stable. There is a tendency towards longer duration at six months post-implant for all of his long vowels except for [i]. AM does not distinguish [a] and [ɔ] on the F1-F2 plane (cf. 4.2.2), but this graph demonstrates a clear length distinction between [a] and [ɔ], which is significant for all recording sessions. There is some sign of change and return, where six month values move towards pre-implant values, particularly when examining [i], [ɔ], and [a]. Statistically significant duration changes, however, are few ([æ] three months vs. six months; [ɔ] pre-implant vs. one month, one month vs. three months, one month vs. six months; [i] one month vs. three months; [ɪ] pre-implant vs. one month).

AM's vowel word duration patterns strongly resemble his vowel durations. The pattern of change and return is evident for [ə], [i], and [ɔ]. The vowel words with statistically significant duration changes are [æ] (pre-implant vs. six months and one month vs. six months); [ɪ] (pre-implant vs.

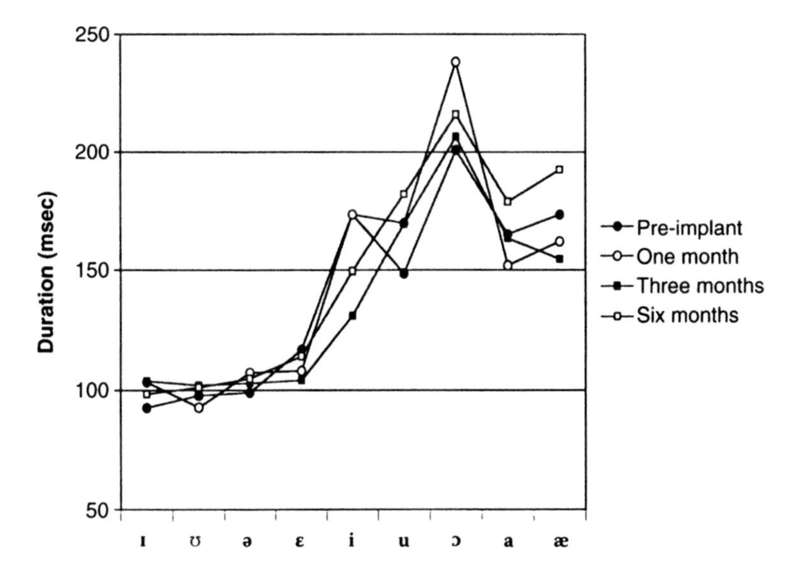

Figure 4.7 Subject AM average vowel duration

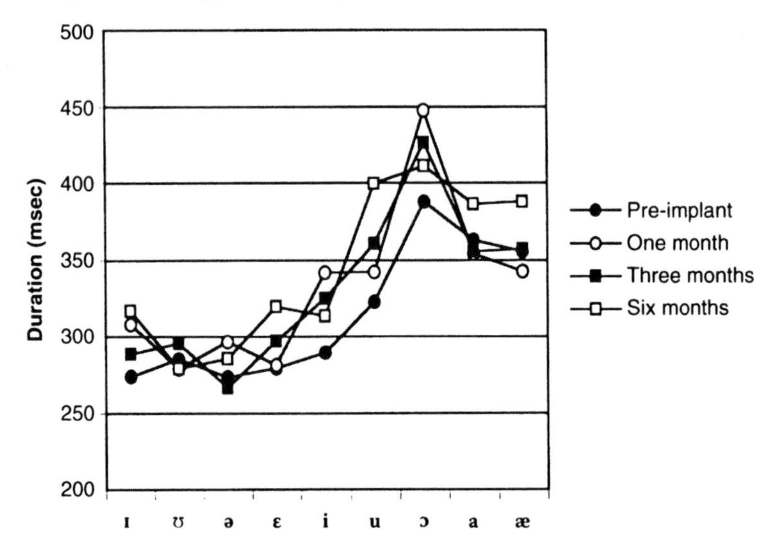

Figure 4.8 Subject AM average vowel word duration

one month); [ə] (one month vs. three months); and [u] (pre-implant vs. three months and pre-implant vs. six months).

For DS, average vowel duration averages 34 milliseconds shorter at 6 months, compared to pre-implant durations. This change is statistically significant for [ɛ], [ɔ] and [u]. In addition, comparing pre-implant and one month and pre-implant and three months, duration changes are significant for [ɛ], [æ], and [u]. Change in vowel duration is significant for all pre to post-implant comparisons for the long vowels when grouped.

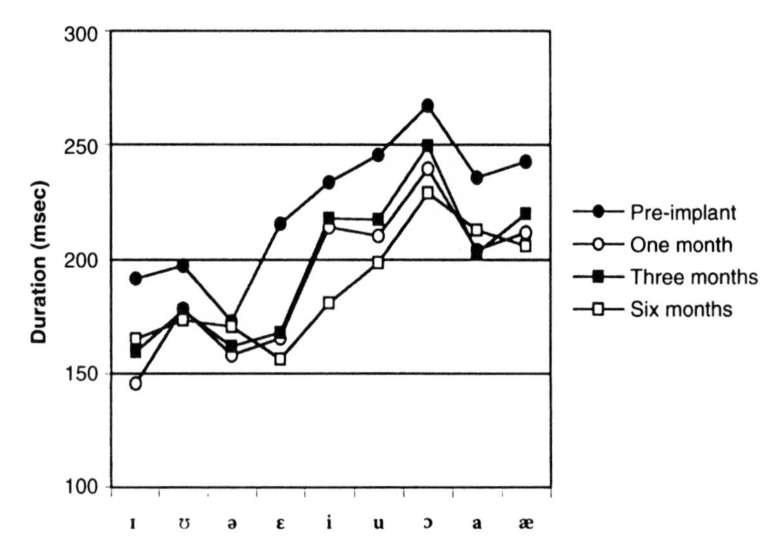

Figure 4.9 Subject DS average vowel duration

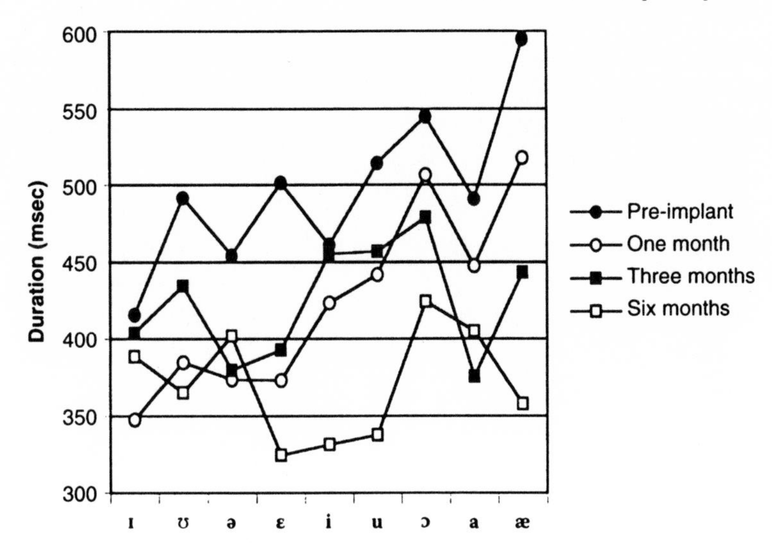

Figure 4.10 Subject DS average vowel word duration

There is a lot of variation in DS's vowel word durations, with no strong trend towards shorter word durations for intrinsically shorter vowels. Her vowel words average 126 milliseconds shorter at six months compared to pre-implant measures; this change is significant for two short vowels ([ʊ], [ɛ]) and most of her long vowels ([u], [ɔ], [a], [æ]). When long vowels are grouped, the change in duration between pre-implant and post-implant is significant.

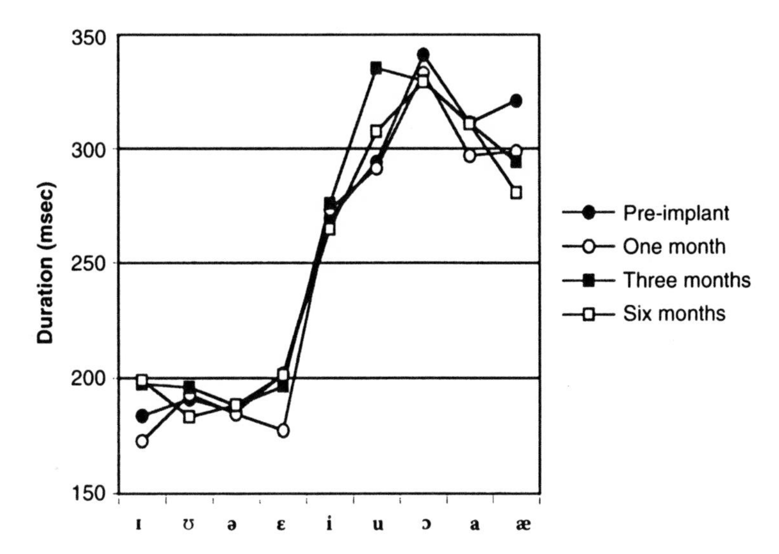

Figure 4.11 Subject MS average vowel duration

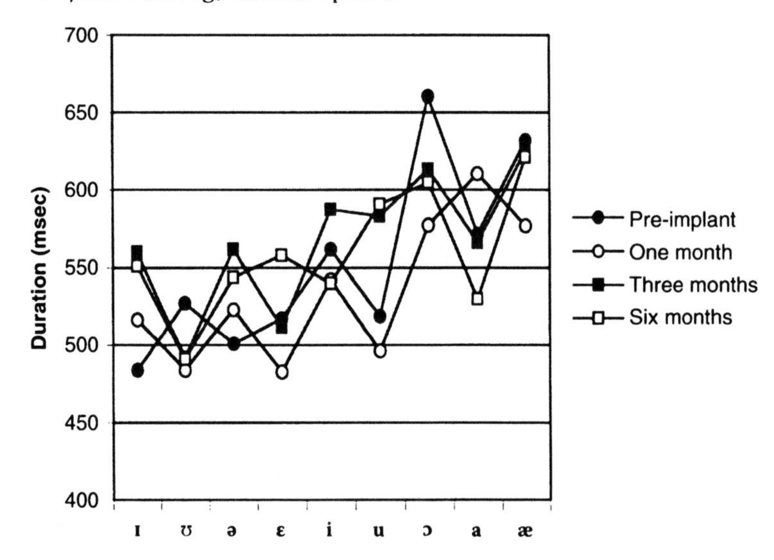

Figure 4.12 Subject MS average vowel word duration

Overall, MS shows the most stability in vowel duration with cochlear implant experience, and there is no overall trend towards either vowel shortening or lengthening; no duration changes are significant when comparing pre-implant and six months. The only significant duration changes are for [ɪ] (one month vs. three month and one month vs. 6 months) and [ɛ] (pre-implant vs. one month and one month vs. six months). A pattern of change and return can be seen for [ʊ], [ɛ], [u], and [a].

MS's strong separation between short and long vowels is not reflected in the durations of her vowel words. She shows a pattern of change and return in vowel word durations for [ɪ], [ə], [i], [ɔ], and [æ]. Only two duration changes are significant: [ɛ] one month vs. six months, and [u] one month vs. three months.

4.3.2 Comparison with Perkell et al. 1992

Subject DS follows the Perkell et al 1992 pattern of decreased vowel duration, but this pattern is only significant for three of her nine vowels. AM's vowel and vowel word durations show a pattern of change and return; MS's data does not demonstrate any strong durational trends.

4.4 VOWEL PEAK FUNDAMENTAL FREQUENCY

4.4.1 Analysis of vowel peak fundamental frequency

The following table consists of the average peak fundamental frequency range for each subject at each recording session, calculated by subtracting the lowest vowel peak F0 from the highest vowel peak F0:

Table 4.4 Average peak fundamental frequency range, all subjects and recording sessions

	AM	*DS*	*MS*
Pre-implant	31 Hz	58 Hz	32 Hz
One month	39 Hz	83 Hz	36 Hz
Three months	46 Hz	48 Hz	51 Hz
Six months	36 Hz	52 Hz	53 Hz

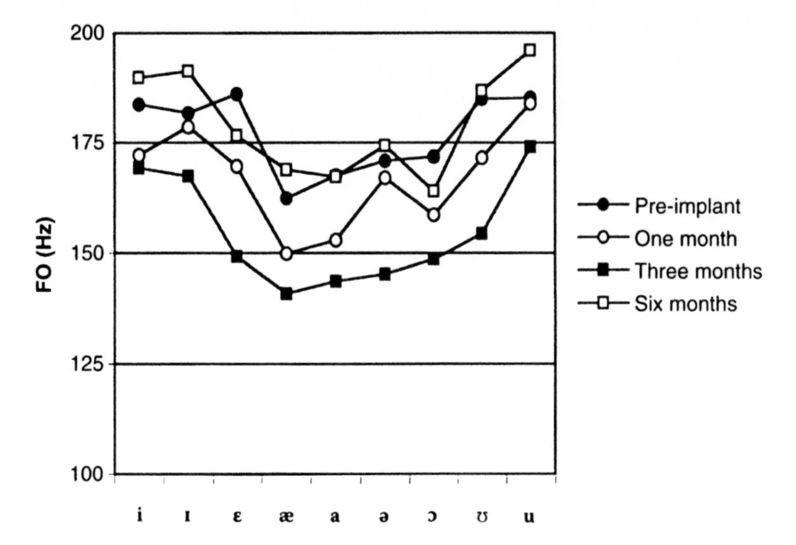

Figure 4.13 Subject AM average vowel peak F0

AM shows very strong evidence of change and return in his peak F0 data. There is a clear drop in F0 at one month and again at three months, whereas the six month values are very close to pre-implant values. This observation is reinforced by statistical analysis; for no vowel is the pre-implant vs. six months comparison significant, and for all vowels but [ɪ] and [æ], at least one of the other comparisons (pre-implant vs. one month, pre-implant vs. three months, one month vs. three months, one month vs. six months, three months vs. six months) is significant. AM's change in average peak F0 range is significant for three months vs. six months.

Rather than a pattern of change and return, DS seems to show less variance with greater cochlear implant experience: compare the wider F0 range and larger F0 differences between vowels at pre-implant and one month with the narrower range and smaller differences at three months and six months. Her peak fundamental frequency overall is higher post-implant. These changes in peak F0 are significant for [æ] (pre-implant vs. one month, pre-implant vs. three months, pre-implant vs. six months), [ɔ] (pre-implant vs. six months, three months vs. six months), [ə] (pre-implant vs. one month), and [ʊ] (pre-implant vs. three months). The changes in peak F0 range are not significant for any comparison.

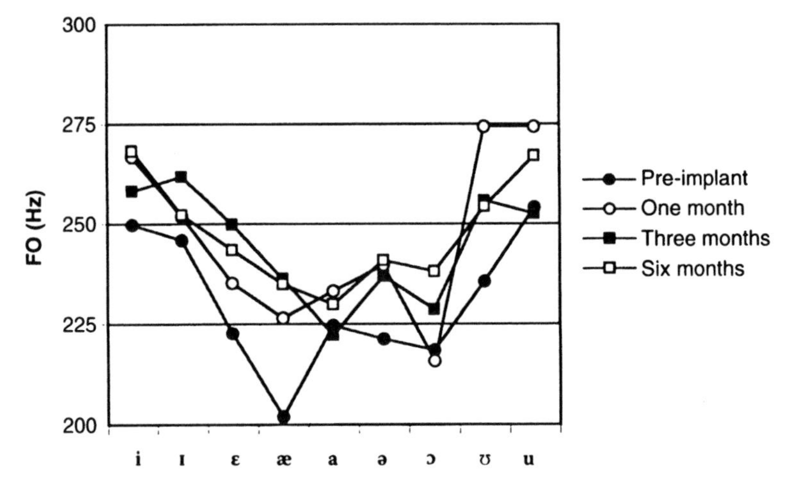

Figure 4.14 Subject DS average vowel peak F0

MS, on the other hand, seems to show more variance in peak F0 with cochlear implant experience, especially at three months, as well as a tendency towards change and return. At pre-implant and one month her peak F0 falls within a range of 32-36 Hz, while at three and six months that range widens to 51-53 Hz. This change in peak F0 range is significant when comparing pre-implant vs. three months, pre-implant vs. six months, one month vs. three months, and one month vs. six months. For all vowels but [ə], [u], and [i], at least one recording session comparison is statistically significant.

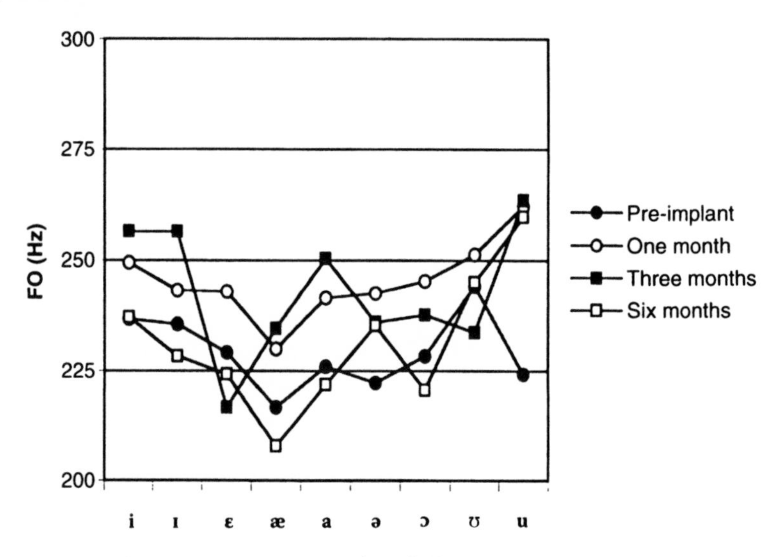

Figure 4.15 Subject MS average vowel peak F0

4.4.2 Comparison with Perkell et al. 1992

Subject DS seems to match Perkell's subjects most closely; she demonstrates fewer F0 differences between vowels with increased cochlear implant experience. MS's results are almost the opposite; she demonstrates larger F0 differences with increased cochlear implant experience. AM's F0 range remains stable, but his data strongly demonstrates change and return, where six month values approach or are closer to pre-implant values than to one month and three month values.

4.5 VOWEL PERCEPTION

4.5. Confusion matrix construction

The confusion matrices used in this section (and in 5.6) were constructed as follows: The results for each subject's perception tests were tabulated, and arranged with the stimulus across the top of the table and the subject's response down the side. Correct responses (the diagonal from top left to bottom right) are indicated in bold.

4.5.2 Vowel perception results

AM perceives the back high vowels [ʊ] and [u] most consistently across sessions. His perception of high front vowels follows an interesting pattern. At one month, he identifies [i] primarily as [u] and [ɪ] primarily as [ʊ]; at three months, his responses change, and he identifies [i] and [ɪ] correctly the majority of the time. At six months, he identifies [i] 18 times out of 18, and [ɪ] 11 times out of 18. [ɛ] follows a similar pattern, with responses of [æ], [ə], [a], and [ʊ] at one and three months changing to a majority of [ɛ] responses at six months. [æ] is initially primarily perceived as [a], but by six months it is perceived as [æ] the majority of the time. AM's perception of [a] follows a different pattern; at one month he identifies [a] correctly 100% of the time, but at three and six months he also identifies [a] as [æ], [ə], and [ʊ]. His perception of [ə] is inconsistent; he identifies [ə] as [ɛ], [æ], [ə], [a], and [ʊ] at each testing session.

It is difficult to find any direct relationships between AM's vowel perception and vowel production measures, but there are some possible factors that relate his perception and production. Examining Figure 4.1, AM's front vowels are higher in F2 (and thus farther from his back vowels) post-implant; this may relate to his one month perception of front vowels primarily as back vowels, and his change to more normal perception of front vowels at three and six months. I note AM's comments that he found himself relying on lipreading more when using the implant; this may also be a factor.

1	i	ɪ	ɛ	æ	ɔ	a	ʊ	u
i	2							
ɪ		2						
ɛ					1			
æ			1	6	2			
ɔ			9	1	8			
a			4	11	4	18	1	
ʊ		16	3		3		17	
u	16							18

One month Stimulus → Response ↓

3	i	ɪ	ɛ	æ	ɔ	a	ʊ	u
i	14							
ɪ	1	10						
ɛ		3	7	4	3		1	
æ			4	11	1	4		
ɔ		1	6		8	2	1	
a			1	3	2	12		
ʊ		4			5		16	
u	3							18

Three months Stimulus → Response ↓

6	i	ɪ	ɛ	æ	ɔ	a	ʊ	u
i	18							
ɪ		11						
ɛ		4	15	6	2			
æ				10	6	5		
ɔ			3	1	5	4	1	
a				1	2	8		
ʊ		3			3	1	17	
u								18

Six months Stimulus → Response ↓

Figure 4.16 Subject AM vowel confusion matrices

1	i	ɪ	ɛ	æ	ɔ	a	ʊ	u
i	18							
ɪ		18						
ɛ			18	16				
æ				2				
ɔ					18	7	1	
a						11		1
ʊ							17	
u								17

One month Stimulus → Response ↓

3	i	ɪ	ɛ	æ	ɔ	a	ʊ	u
i	18							
ɪ		18						
ɛ			11					
æ			7	18				
ɔ					18		2	
a						18		
ʊ							16	1
u								17

Three months Stimulus → Response ↓

6	i	ɪ	ɛ	æ	ɔ	a	ʊ	u
i	18							
ɪ		17						
ɛ		1	2					
æ			16	18				
ɔ					17			
a					1	18		
ʊ							18	
u								18

Six months Stimulus → Response ↓

Figure 4.17 Subject DS vowel confusion matrices

DS has remarkably consistent vowel perception, with three exceptions: [a], [ɛ], and [æ]. At one month, she identifies [a] as [ə] one third of the time, but at three and six months she perceives [a] consistently. The perception test speaker's [ɛ] and [æ] seem to fall into the same perceptual category for her dialect; she identifies [ɛ] and [æ] both as [ɛ] at one month, at three months she identifies [ɛ] correctly 11 times out of 18 and [æ] perfectly, while at six months, she identifies [ɛ] and [æ] both as [æ].

Comparing DS's vowel perception with her vowel production, there are no direct correlations between her single-category perception of [ɛ] and [æ] and her production of [ɛ] and [æ] (cf. Figure 4.3). Her one month perception of [a] as [ə] might be related to how close those two vowels are to each other in the F1-F2 plane at one month.

At one month, MS's perception of high front vowels follows a similar pattern to that of AM, though to less of a degree: [i] is identified as [u] some of the time, and [ɪ] is identified as [ʊ] (or [u]) most of the time. Her [i] perception normalizes at three months, while her [ɪ] perception improves until at six months, she is identifying [ɪ] correctly 15 times out of 18. Her perception of [a], [ʊ], and [u] is consistently good across testing sessions. Like DS, another Midwesterner, MS seems to put the perception test speaker's [ɛ] and [æ] into the same perceptual category; at one month, she is identifying [ɛ] and [æ] correctly the majority of the time, but at three and six months, both [ɛ] and [æ] are primarily identified as [æ]. She identifies [ə] primarily as [a] across sessions, but also as [ə] and [ʊ].

1	i	ɪ	ɛ	æ	ə	a	ʊ	u
i	**12**							
ɪ	2	**3**						
ɛ		3	**10**	**5**				
æ			7	**13**				
ə			1		**9**			
a					6	**18**		
ʊ		9			3		**18**	
u	4	2						**18**

One month Stimulus → Response ↓

3	i	ɪ	ɛ	æ	ə	a	ʊ	u
i	**18**	2						
ɪ		**8**						
ɛ		1	**4**	2				
æ			14	**16**				
ə					**6**		2	
a					11	**18**		
ʊ		3			1		**15**	1
u		3					1	**17**

Three months Stimulus → Response ↓

6	i	ɪ	ɛ	æ	ə	a	ʊ	u
i	**18**							
ɪ		**15**						
ɛ			**6**					
æ			12	**18**				
ə					**8**			
a					9	**18**		
ʊ		3			1		**18**	
u								**18**

Six months Stimulus → Response ↓

Figure 4.18 Subject MS vowel confusion matrices

MS's perception of [ɛ] as [æ] may be related to the diphthongized nature of her [æ]; the first half of her diphthong is close to her production of [ɛ] (cf. Figure 4.5). Her perception of [ə] as [a] may be related to the F2 similarity of her productions of [ə] and [a].

4.6 DISCUSSION

In terms of vowel production measures, DS is most similar to the subjects in Perkell et al. 1992. With cochlear implant experience, her F2s shift in a normative direction, and her vowel space expands; her vowel duration and vowel word duration decreases; and there are fewer F0 differences between her vowels.

AM, on the other hand, shows a consistent pattern of change at one and three months and return to pre-implant levels at six month for his vowel production measures. His front vowels shift upwards in F2, but his back vowels shift as well, and his vowel space (as characterized by both AVS and triangle area) at six months is nearly the same as at pre-implant. AM's vowel and vowel word durations demonstrate the same pattern of change and return for his long vowels. His peak F0 data also strongly illustrates change and return, and the F0 differences between his vowels are fairly stable across recording sessions.

MS's overall vowel space (as characterized by AVS) shows a pattern of change and return; her vowel durations are stable, but more than half of her vowel word durations demonstrate change and return. Unlike the Perkell subjects, MS exhibits a greater degree of F0 differences between vowels with increased cochlear implant experience.

In terms of vowel perception, DS and MS (both from the Midwest) categorize the hearing speaker's [ɛ] and [æ] as [æ] at six months post-activation. AM and MS both perceive high front vowels as high back vowels at the one month testing session. This may be due to initial difficulty in perceiving high F2s—they seem to be classifying high front vowels as back using F1. DS, on the other hand, perceives high front vowels correctly at all sessions.

5 Stops and consonant perception

5.1 PREVIOUS FINDINGS

Lane and Webster (1991) found that their postlingually deafened subjects showed a lesser degree of place differentiation in stop consonants. The VOT study by Lane, Wozniak, and Perkell (1994) found that for all of their subjects, CV syllable duration became significantly shorter after implant activation, so that it was impossible to compare pre- and post-activation VOT without using a correction factor (VOTc). Three of their four speakers lengthened VOTc post-activation. In addition, they found that VOT place distinctions did not change with CI experience. See **2.4.5** for a detailed analysis of this study.

5.2 CV SYLLABLE DURATION

AM's average CV syllable duration tends to be longer post-activation, and this change is significant for both voiced (pre-implant vs. three months,

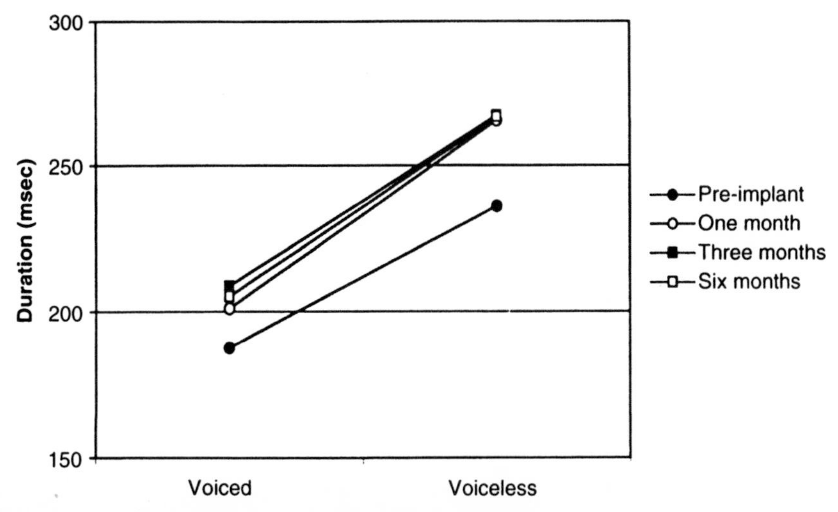

Figure 5.1 Subject AM average CV syllable duration

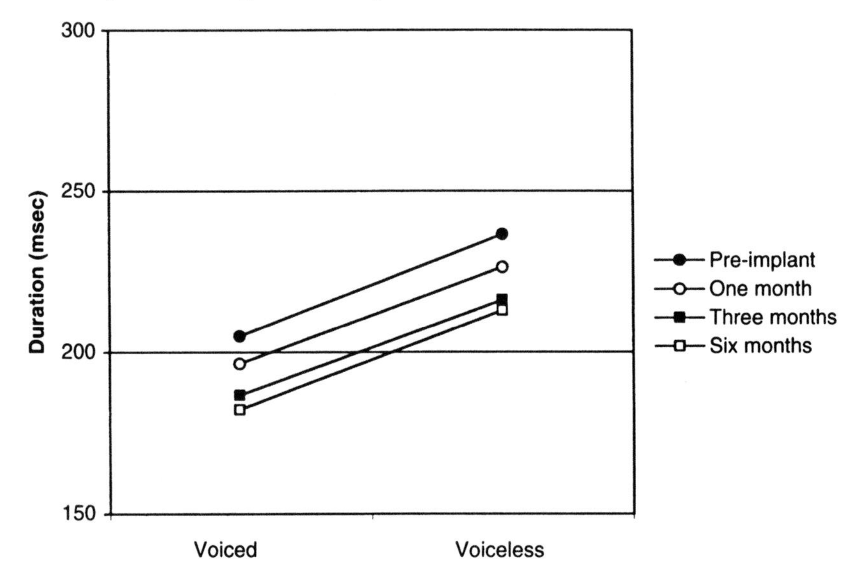

Figure 5.2 Subject DS average CV syllable duration

pre-implant vs. six months) and voiceless (pre-implant vs. one month, pre-implant vs. three months, pre-implant vs. six months) CV syllables.

For DS, average CV syllable duration becomes shorter with cochlear implant experience. This change is not significant for voiced or voiceless CV syllables, but it is significant for pre-implant vs. three months and pre-implant vs. six months when all of her CV syllable data is grouped.

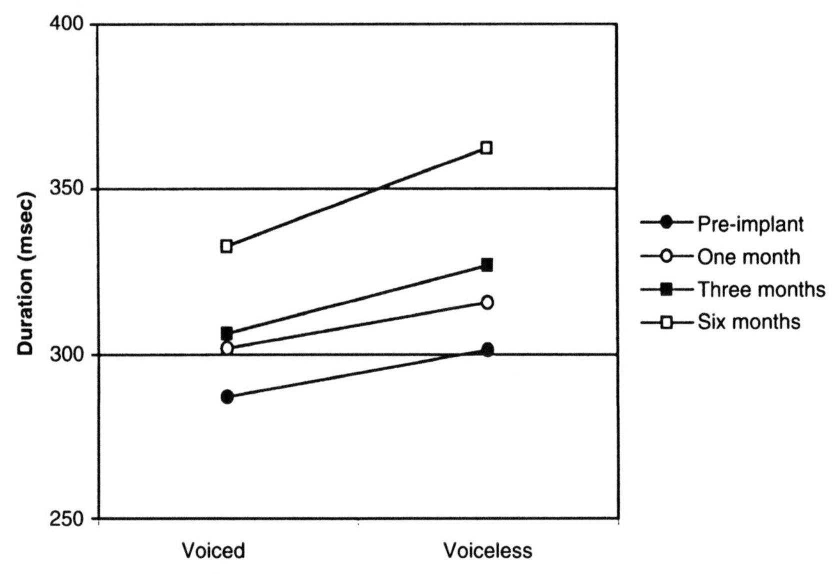

Figure 5.3 Subject MS average CV syllable duration

MS's average CV syllable duration is longer post-activation, and this change is significant for pre-implant vs. six months for both voiced and voiceless CV syllables. In addition, this change is significant for voiceless CV syllables for one month vs. six months and three months vs. six months.

5.3 COMPARISON WITH LANE ET AL. 1994

Like the subjects in Lane et al. 1994, overall CV syllable duration did change significantly with cochlear implant experience. The question is whether it is necessary to use a correction factor when comparing VOT from pre- and post-activation recording sessions.

VOTc was calculated using the metric proposed in Lane et al. 1994:

> The mean syllable durations of voiced and of voiceless tokens in the first baseline session were computed separately for each speaker. Then, for each of a speaker's voiced and voiceless tokens in all the other sessions, the appropriate mean from the first baseline session (voiced or voiceless) was subtracted from the syllable duration of the token to obtain its change in duration from baseline...The change in VOT associated with this change in syllable duration was obtained...by multiplying the syllable duration by the appropriate slope constant, 0.021 for voiced tokens and 0.132 for voiceless tokens...Finally, this change in VOT...was subtracted from the VOT of the token, to obtain...VOTc (59).

Table 5.1 displays average VOT and VOTc for each subject at the six month recording session. The difference between VOT and VOTc for each subject was only significant for subject MS, /k/, six months (p=.029). Given the overall lack of statistical significance, I do not believe it is necessary to use VOTc to accurately compare pre-implant VOTs to post-implant VOTs.

Table 5.1 Average VOT and VOTc for all subjects, six month recording session

(=p<.05)*	AM		DS		MS	
6 months	**VOT(msec)**	**VOTc(msec)**	**VOT(msec)**	**VOTc(msec)**	**VOT(msec)**	**VOTc(msec)**
/b/ avg	−1.96	−2.02	0.81	1.66	5.56	5.21
/d/ avg	1.67	1.32	2.98	3.53	10.84	9.71
/g/ avg	5.22	4.53	2.08	2.12	7.12	5.73
/p/ avg	100.11	98.86	52.30	55.87	48.82	42.82
/t/ avg	120.43	115.65	58.30	63.83	66.71	58.43
/k/ avg	124.68	118.65	71.32	71.55	62.61	52.81*

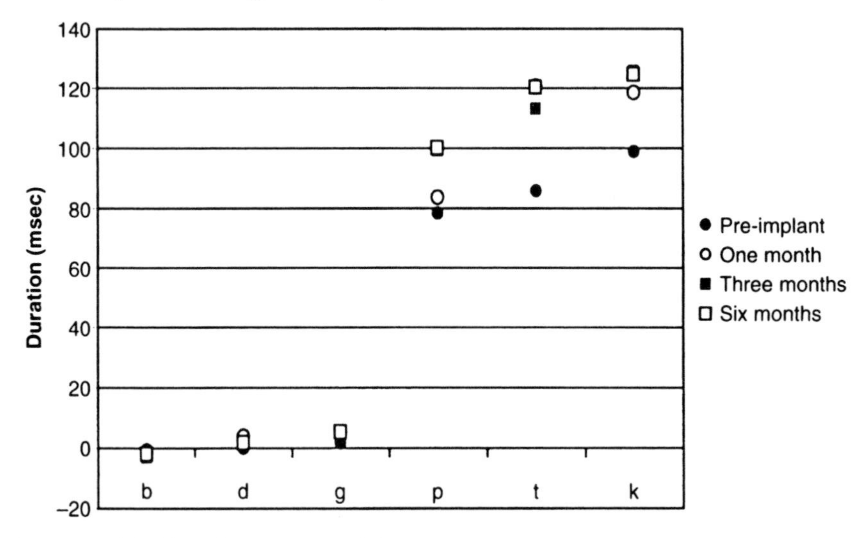

Figure 5.4 Subject AM average VOT

5.4 VOT

Overall, AM shows an increase in VOT post-activation. This increase is significant for all voiceless stops when comparing pre-implant and six months, while the changes in VOT for voiced stops are not significant for any time comparison. AM has a clear VOT distinction between voiced and voiceless stops, and also a VOT distinction between [p] and the other two voiceless stops ([p] vs. [t] is significant for all sessions but pre-implant; [p] vs. [k] is significant for all sessions). [t] and [k] are only significantly distinguished by VOT at the three month recording session. AM does not produce significant VOT distinctions between his voiced stops.

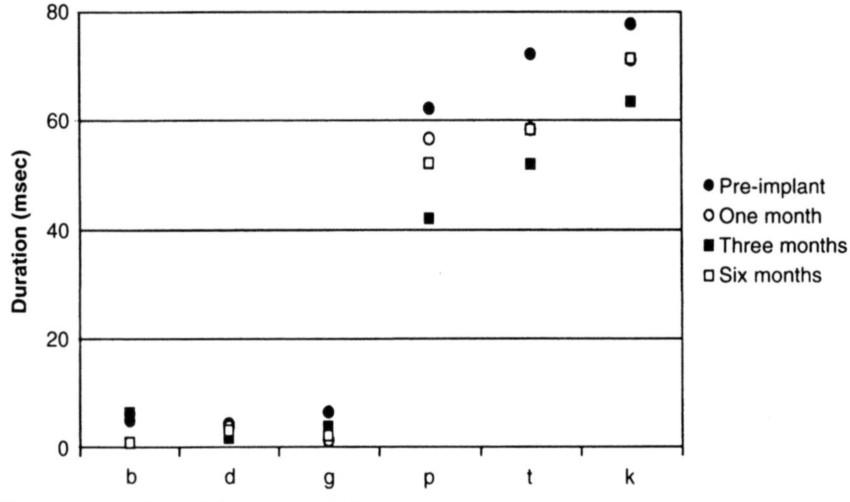

Figure 5.5 Subject DS average VOT

DS shows decreased VOTs with cochlear implant experience, with her lowest VOTs at three months and her six month VOTs very close to her one month VOTs—a similar pattern of change and return to that found in the vowel data. Changes in VOT are significant for [p] and [t] (pre-implant vs. three months and pre-implant vs. six months) and also for [b] (one month vs. six months, three months vs. six months) and [g] (pre-implant vs. one month). DS makes a clear voiced-voiceless VOT distinction, and appears to use VOT to distinguish place for voiceless stops. [p] and [k], and [t] and [k], have significant VOT differences at three and six months; [p] and [t] are not significantly distinguished by VOT at any session. DS does not produce significant VOT distinctions between her voiced stops.

MS shows an interesting overall pattern in VOT changes for her voiceless stops; a drop in VOT at one month, three months is very close to pre-implant, and six months is higher than pre-implant. These changes are significant for [t] for all comparisons but pre-implant vs. three months, and for [k] for pre-implant vs. one month, one month vs. three months, one month vs. six months, and three months vs. six months. The voiced stops do not follow any general trend; changes in [b] are significant when comparing one month and three months, changes in [d] are significant for pre-implant vs. six months and one month vs. six months, while there are no significant changes in [g]. MS makes a clear voiced-voiceless distinction, and appears to use VOT to distinguish [p] from [t] and [k] (significant for pre-implant, three months, and six months). It is interesting that that distinction is not maintained at one month. MS does not use VOT to distinguish [t] and [k]; note that MS consistently produces [k] with a shorter VOT than [t]. MS does not produce significant VOT distinctions between her voiced stops.

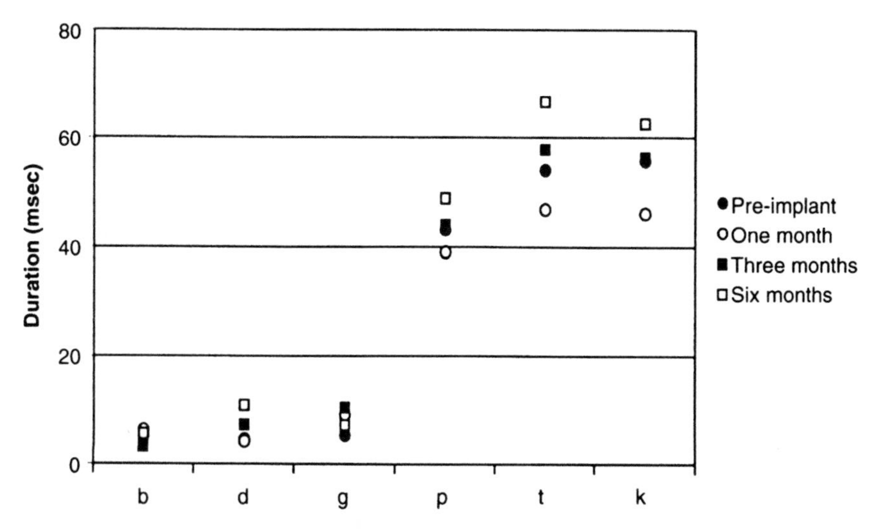

Figure 5.6 Subject MS average VOT

5.5 COMPARISON WITH LANE ET AL. 1994

Lane et al. found that for three of their four speakers, there was an increase in VOTc post-activation. AM and MS show increases in VOT for voiceless stops (comparing pre-implant and six months), while DS has shorter VOTs. The voiced-voiceless distinction is stable for all subjects when comparing pre-implant and post-activation data, and each subject maintains a significant distinction between [p] and [k].

5.6 CONSONANT PERCEPTION

AM's perception of [b], [p], [s], and [ʃ] is the most consistent over all testing sessions. The alveolar stops [d] and [t] are the next in terms of perceptive consistency, with 13-17/18 perceived correctly at each session. When confusions occur, AM tends to identify [d] as [g] and [t] as [k] (except at one month, where he identified [t] as [p]). The changes in velar perception are the most interesting; at one month, AM identifies the majority of [g]s as [d]s, and [k]s as [t]s. At three months, his responses to the velar stimuli are evenly split between velar and alveolar responses; at six months, he is identifying [g] primarily as [d] (note that there are more [g] responses than at one month) but is identifying [k] correctly more than 50% of the time. AM's perceptual confusions seem to be based on place rather than voicing.

1	b	p	d	t	g	k	s	ʃ
b	17							
p		15		3		1		
d			17		16			
t		2		15		12		
g	1		1		2			
k		1				4		
s						1	18	
ʃ								18

One month Stimulus → Response ↓

3	b	p	d	t	g	k	s	ʃ
b	16							
p	1	18	1					
d	1		14		9			
t				13		8		
g			3	1	9			
k				4		10		
s							18	
ʃ								18

Three months Stimulus → Response ↓

6	b	p	d	t	g	k	s	ʃ
b	16				1			
p	1	18	2					
d			14		11			
t				13		7		
g	1		2		6			
k				5		11		
s							18	
ʃ								18

Six months Stimulus → Response ↓

Figure 5.7 Subject AM consonant confusion matrices

1	b	p	d	t	g	k	s	ʃ
b	8							
p		15						
d	10		18					
t		3		18				
g					18	1		
k						17		
s							18	
ʃ								18

One month Stimulus → Response ↓

3	b	p	d	t	g	k	s	ʃ
b	14							
p		17						
d	4		18		1			
t		1		17				
g				1	17	4		
k						14		
s							18	
ʃ								18

Three months Stimulus → Response ↓

6	b	p	d	t	g	k	s	ʃ
b	15							
p		18						
d	2		18					
t				18				
g	1				18	5		
k						13		
s							18	
ʃ								18

Six months Stimulus → Response ↓

Figure 5.8 Subject DS consonant confusion matrices

AM does not produce a significant VOT distinction between [g] and [d] at any session; this may be reflected in his [g]-[d] identification errors. The same is true for [k] and [t], which are not produced with a significant VOT distinction (except for at three months). [p] and [t] are produced with a significant VOT distinction post-activation but not pre-implant; AM's minor [t]-[p] confusions at one month may reflect this.

DS identifies [d], [t], [g], [s], and [ʃ] consistently across testing sessions. At three and six months, she sometimes identifies [k] as [g]. At one month post-activation, she identifies [b] as [d] slightly more than [b]; at three and six months she still identifies [b] as [d] some of the time, but identifies it correctly most of the time. DS's [k] confusions seem to be based on voicing, and her [b] ones on place of articulation.

DS produces [k] and [g] with distinctive VOTs ($p<.00001$); her [k]-[g] errors can't be explained as a function of VOT. She does not produce a significant VOT difference between [b] and [d] at any session; this may be reflected in her identification of [b] as [d], particularly at the one month session.

At one month, MS scores near 100% on consonant perception, with only a few errors; at three months, she does score 100%. At six months, she identifies [t] as [k] five times out of 18, a confusion likely based on place of articulation; otherwise, she identifies these consonants at or near 100%.

MS does not produce a significant VOT distinction between [t] and [k] at any session; this may be reflected in her identification of [t] as [k] at six months.

1	b	p	d	t	g	k	s	ʃ
b	16							
p		18						
d	2		18		1			
t				17				
g					17	1		
k				1		17		
s							18	
ʃ								18

One month Stimulus → Response ↓

3	b	p	d	t	g	k	s	ʃ
b	18							
p		18						
d			18					
t				18				
g					18			
k						18		
s							18	
ʃ								18

Three months Stimulus → Response ↓

6	b	p	d	t	g	k	s	ʃ
b	17							
p		18						
d			18					
t				13				
g					18			
k				5		18		
s	1						18	
ʃ								18

Six months Stimulus → Response ↓

Figure 5.9 Subject MS consonant confusion matrices

5.7 DISCUSSION

All three subjects in this study changed their CV syllable duration significantly by three months post-activation, though only DS's durations became shorter (which is most like the Lane et al. 1994 subjects). Unlike the Lane et al. subjects, CV syllable duration did not change to the extent that the use of VOTc was necessary. AM and MS both increased their voiceless VOTs post-activation, while DS's voiceless VOTs decreased.

In terms of consonant perception, all three subjects perceive the /s/-/ʃ/ contrast at all testing sessions, perhaps indicating the greater degree of high frequency information provided by modern cochlear implant processors. DS and MS exhibit consistently good stop perception at all testing sessions, with the exception of DS's [b]-[d] place confusions at one month. AM exhibits a greater degree of perception errors across sessions—even at six months post-activation, he is not consistently distinguishing velars from alveolars (though he has improved from his performance at one month).

6 Intonation

6.1 PREVIOUS FINDINGS

Lane and Webster (1991) found that their postlingually deafened subjects produced greater variations in pitch than comparable hearing subjects, and Rubin-Spitz and McGarr (1990) found their deaf subjects made little F0 differentiation between declarative and nondeclarative sentences. Lane et al. 1997, which uses the Rainbow Passage, found that three of their four cochlear implant subjects produced less variable F0 contours post-activation; the fourth subject maintained the same level of contour variability. See **2.4.5** for a detailed analysis of this study.

6.2 THE RAINBOW PASSAGE

6.2.1 Calculation of contour variability

Contour variability was calculated using the method outlined in Lane et al. 1997. Each subject's data was analyzed individually. For each phrase of the Rainbow Passage (see **3.2.3.3** for phrase divisions), the average F0 of each syllable in the phrase was divided by the highest average syllable F0 and expressed as a percent (2247). For details on how average syllable F0 was calculated, see **3.2.3.3**. Then, mean successive difference (MSD) was calculated as follows:

> One way of parametrizing the variation…is to subtract the percent-of-maximum-value of each syllable from that of the following syllable. These successive differences are positive when the contour is rising and negative where it is falling; the average of their absolute values is the mean successive difference (MSD) (2247).

MSD values for each sub-phrase were tested for statistical significance, comparing each recording session.

Sample calculation:

Table 6.1 Sample calculation of MSD

MSrp05	*with*	*its*	*path*	*high*	*a*	*bove*	
pre 1	205.343434	198.895914	203.895227	198.4609	189.464821	209.860437	
pre 2	221.446147	206.381124	199.66778	194.555812	188.188232	171.160766	
pre 3	212.080165	198.582549	197.320685	189.103018	184.0815	186.790238	
	with %	*its %*	*path %*	*high %*	*a %*	*bove %*	*MSD*
pre 1	97.8476158	94.7753261	97.1575346	94.5680391	90.2813428	100	4.40986945
pre 2	100	93.1969813	90.165389	87.8569419	84.9814887	77.2922756	4.54154488
pre 3	100	93.6356065	93.0406127	89.1658197	86.7980748	88.0752984	2.89582977

6.2.2 Rainbow Passage figures

The Rainbow Passage figures referred to in this chapter are contained in **Appendix C.**

6.2.3 Rainbow Passage analysis

AM has the highest MSD values, ranging from 9.88% to 11.33%. Looking at his MSD values across all phrases, the pre-implant vs. three months change is significant and the three months vs. six months change is highly significant ($p<.005$).

Table 6.2 MSD comparisons

MSD	*AM*	*DS*	*MS*
Pre-implant	10.78 %	7.55 %	4.96 %
One month	10.55 %	9.70 %	4.88 %
Three months	9.88 %	9.0 %	6.04 %
Six months	11.33 %	9.93 %	5.47 %

Examining AM's fundamental frequency patterns for each Rainbow Passage phrase, a tendency towards change and return emerges. F0 tends to be lower than pre-implant levels at one and three months, and returns to pre-implant levels at six months. This is particularly true for phrase 1 (**Figure C.1**), phrase 7 (**Figure C.7**) and phrase 10 (**Figure C.10**). Another set of phrases follow the change and return pattern at the start of the phrase, but six month F0s are higher at the end of the phrase (phrase 2 (**Figure C.2**), phrase 3 (**Figure C.3**), phrase 4 (**Figure C.4**), phrase 6 (**Figure C.6**), phrase 11 (**Figure C.11**)). Six month F0s are higher than pre-implant F0s for phrases 5 (**Figure C.5**), 6 (**Figure C.6**), 8 (**Figure C.8**), and 9 (**Figure C.9**). AM is fairly consistent in his placement of frequency peaks in all phrases across recording sessions.

AM's increase in Rainbow Passage total duration is significant for pre-implant vs. one month, pre-implant vs. three months, pre-implant vs. six months, and one month vs. six months. Examining the duration changes of individual phrases, the duration changes of phrases 5, 6, 9, and 11 are significant for pre-implant vs. six months.

Table 6.3 AM Rainbow Passage total duration (average)

AM	*Duration*
Pre-implant	31.77 sec
One month	34.11 sec
Three months	35.56 sec
Six months	37.54 sec

DS's MSD values (**Table 6.2**) range from 7.55% to 9.93%. Looking across her Rainbow Passage phrases for all sessions, the pre-implant vs. one month and pre-implant vs. six months comparisons are highly significant ($p<.005$).

Looking at changes in average syllable F0 over time, in phrase 1 (**Figure C.12**), DS produces lower F0s at the start of the phrase and higher ones at the end of the phrase, comparing pre-implant and post-implant sessions. In phrases 6, 8, 9, 10, and 11 (**Figure C.17, Figure C.19, Figure C.20, Figure C.21, Figure C.22**), she produces higher F0s at the start of the phrase and lower ones at the end of the phrase, comparing pre-implant and post-implant sessions. In phrases 2, 3, 4, 5, and 10 (**Figure C.13, Figure C.14, Figure C.15, Figure C.16, Figure C.21**), the syllable with the highest frequency changes between pre-implant and post-implant. Phrases 3, 4, and 10 (**Figure C.14, Figure C.15, Figure C.21**) also reveal that DS used a different melody at the start of the phrase at each recording session. In phrases 5 and 7 (**Figure C.16, Figure C.18**), DS produces a different overall F0 pattern at each recording session.

Table 6.4 DS Rainbow Passage total duration (average)

DS	*Duration*
Pre-implant	38.37 sec
One month	33.35 sec
Three months	31.81 sec
Six months	29.23 sec

DS's reduction in total Rainbow Passage duration is highly significant for pre-implant vs. six months, one month vs. six months, and three months vs. six months ($p<.005$), and also significant for pre-implant vs. one month and pre-implant vs. three months. Examining the duration changes of individual phrases, most pre-implant vs. post implant comparisons are significant, and pre-implant vs. six months is significant for all phrases.

MS's MSD values (**Table 6.2**) hover around 5%. Looking at her Rainbow Passage phrases across all recording sessions, the increase in MSD between one month and three months is significant.

In terms of F0 changes, for individual Rainbow Passage phrases, MS primarily produces higher F0s at one month, then the three and six month F0s are lower than pre-implant values. This pattern can be seen in phrases 1, 2, 3, 4, 5, 6, 8, 9, 10, and 11 (**Figure C.23, Figure C.24, Figure C.25, Figure C.26, Figure C.27, Figure C.28, Figure C.30, Figure C.31, Figure C.32, Figure C.33**). Phrase 7 (**Figure C.29**) is the only case where the primary F0 pattern is not followed—here, the post-implant F0s are overall higher than pre-implant values. In addition, for some phrases, the syllable with the highest frequency changes from pre- to post-implant. This is true for phrase 1 (**Figure C.23**), phrase 3 (**Figure C.25**), phrase 7 (**Figure C.29**), phrase 10 (**Figure C.32**), and phrase 11 (**Figure C.33**).

Table 6.5 MS Rainbow Passage total duration (average)

MS	*Duration*
Pre-implant	33.45 sec
One month	34.43 sec
Three months	33.25 sec
Six months	32.61 sec

MS's total Rainbow Passage duration does not change significantly with increased cochlear implant experience. Examining the duration changes of individual phrases, the change in duration from pre-implant to six months is significant only for phrase 5.

6.3 BEV LOVES BOB SENTENCES

6.3.1 Bev loves Bob figures

The Bev loves Bob figures are located in **Appendix D**.

6.3.2 Analysis

I attempted to parameterize the variation in Bev loves Bob peak F0s in two ways: calculating MSD (**6.2.1**), and calculating average frequency of the sentence (following Rubin-Spitz and McGarr 1990). Neither of these methods sufficiently captured the overall patterns seen in the graphs, so this section is based on visual observations of the averaged and unaveraged Bev loves Bob peak F0 graphs.

AM shows a consistent pattern of change and return across his Bev loves Bob utterances, where one and three month peak F0s are lower than pre-implant values, and six month peak F0s are close to or higher than pre-implant values. This is true for "Bev loves Bob." (**Figure D.1**), "<u>Bev</u> loves Bob." (**Figure D.3**), "Bev <u>loves</u> Bob." (**Figure D.5**), "Bev loves <u>Bob</u>." (**Figure D.7**), "Bev loves Bob?" (**Figure D.9**), and "Bev loves <u>Bob</u>?" (**Figure D.15**). In "<u>Bev</u> loves Bob?" (**Figure D.9**) and "Bev <u>loves</u> Bob?" (**Figure D.11**), the 6 month peak F0s are lower than pre-implant values, but otherwise the change and return pattern holds.

For the most part, the average graphs present an accurate picture of AM's peak F0 patterns, but there are two exceptions: "<u>Bev</u> loves Bob." (**Figure D.4**), and "Bev loves Bob?"(**Figure D.10**). In the case of "<u>Bev</u> loves Bob.", one pre-implant utterance has a much higher F0 for "loves" and "Bob", which raises the average pre-implant F0. In the case of "Bev loves Bob?" there is a lot of variation in AM's pre-implant utterances, which is lost when averaging the data.

DS tends to produce higher F0s post-implant; in terms of intonation patterns, however, there are no strong overall trends in DS's Bev loves Bob data. In "Bev loves Bob." (**Figure D.18**) and "Bev loves Bob" (**Figure D.24**) she produces the most varied intonation at pre-implant and three months, and the one and six month F0 patterns are less varied. For these phrases, the average F0 graph (**Figure D.17, Figure D.23**) does not accurately portray the F0 variation. In "Bev loves Bob." (**Figure D.19**) and "Bev loves Bob?" (**Figure D.25**) DS uses a less extreme F0 range with greater cochlear implant experience. In "Bev loves Bob." (**Figure D.22**) and "Bev loves Bob?" (**Figure D.30**) DS uses a more extreme F0 range with greater cochlear implant experience; in addition, her three month productions show the greatest degree of variation, and the average F0 graphs (**Figure D.21, Figure D.29**) do not present the full picture of her F0 patterns. In "Bev loves Bob?" (**Figure D.28**) and "Bev loves Bob" (**Figure D.32**) the widest F0 range occurs in the three month session; otherwise, DS produces the sentence fairly consistently across recording sessions.

MS tends to produce Bev loves Bob sentences with more variability post-activation, and has a tendency to produce the most variation at one month. This is true for "Bev loves Bob." (**Figure D.40**), "Bev loves Bob?" (**Figure D.42**), "Bev loves Bob?" (**Figure D.44**), and "Bev loves Bob?" (**Figure D.48**). "Bev loves Bob." (**Figure D.34**) has the highest degree of variation at three months, while "Bev loves Bob." (**Figure D.36**) has the highest degree of variation at six months. Her productions of "Bev loves Bob?" (**Figure D.46**) are highly variable at every session. Her productions of "Bev loves Bob." (**Figure D.38**) are the only ones that are produced consistently across sessions. The average graph (**Figure D.37**) is accurate for this data set, but overall, the average graphs do not portray the variability of MS's productions. I note that MS's Bev loves Bob sentences post-activation were very difficult to categorize by ear, and that MS commented at the post-implant recording sessions that that these intonation patterns were difficult for her to produce, as she found she can easily hear intonation in the speech of others, but not in her own speech.

6.4 DISCUSSION

Unlike the subjects in Lane et al. 1997, the three subjects in this study all produce somewhat greater F0 contour variation (as captured using MSD) in their Rainbow Passage phrases post-activation. AM continues to exhibit a tendency towards change and return, particularly in his F0 patterns. DS significantly decreases Rainbow Passage phrase duration post-implant, and tends to change which syllable has the most prominent frequency from pre-implant to post-activation. MS's Rainbow Passage phrase duration does not change with cochlear implant experience, but she does tend to produce lower F0s at three and six months than she does in the pre-implant recording sessions. Like DS, MS tends to change which syllable is the most prominent post-implant.

The Bev loves Bob sentences are more difficult to categorize. For AM and DS, averaging across repetitions generally captured their intonation patterns, but for MS, her individual productions vary to an extent that average graphs cannot capture. AM's Bev loves Bob data show a consistent pattern of change and return. I do not observe any general trends in DS's Bev loves Bob intonation patterns.

7 Hearing subjects' perception of cochlear implant users speech

7.1 BACKGROUND

The acoustic changes in the cochlear implant subjects' speech in this study are often subtle; therefore, it is interesting to study whether naïve listeners can distinguish speech produced before cochlear implant surgery from speech produced after six months of cochlear implant experience.

7.2 METHODOLOGY

7.2.1 Subjects

Subjects for this study consisted of eighteen native speakers of English (9 male, 9 female) who had normal hearing, no experience with deaf or hearing impaired speech, and were students at the University of Chicago. The mean naïve subject age was 21.6 (median 21). In addition, an expert listener (an adult native speaker who is a speech and hearing therapist) participated in this study.

7.2.2 Materials

Recordings of the Rainbow Passage from each of the CI users' pre-implant and six month recording sessions were digitized at 20 kHz (MS) or 48 kHz (DS, AM) and split into eleven phrases by breath group (see 3.2.3.3). Each subject block (CI block) consisted of 132 randomized stimuli (2 repetitions X 3 presentations X 11 phrases X 2 recording sessions).

7.2.3 Procedures

Subjects were first presented with the text of the Rainbow Passage onscreen and then instructed to decide as quickly and accurately as possible whether a particular phrase came from the pre-implant or post-implant recording session. A short block consisting of the eleven phrases of the Rainbow

Passage, in order, with samples from each speaker from most recording sessions, was presented before the CI blocks. The three CI blocks were presented in all six possible orders (three subjects assigned to each order) using Inquisit, and the subjects answered by pressing one of two keys on the keyboard. Answers and latency (response time, in milliseconds) were recorded.

7.2.4 Analysis

An overall correctness score and average latency was generated for each CI speaker and recording session for each hearing subject. These results were averaged across subject and subjected to ANOVA (Analysis of Variation) using StatView.

7.3 RESULTS

7.3.1 Naïve listeners

Examining average correctness, it appears that it is easier for naïve subjects to identify post-implant speech than it is for them to identify pre-implant speech. It is also clear that subject AM's speech is the most difficult to distinguish; the identification score for his pre-implant speech is 34% and the score for his post-implant speech is 55%—near chance. DS's post-implant

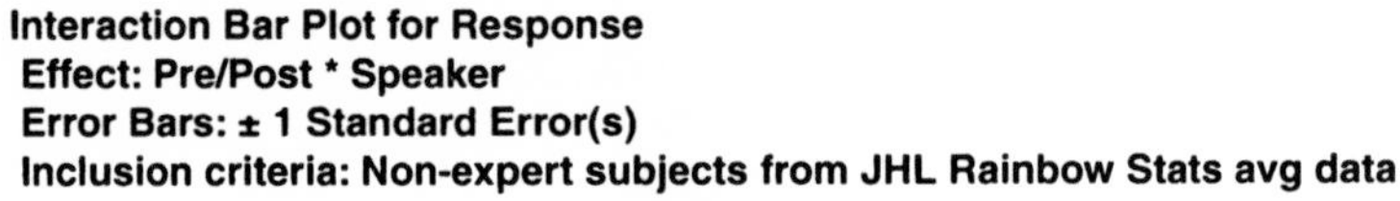

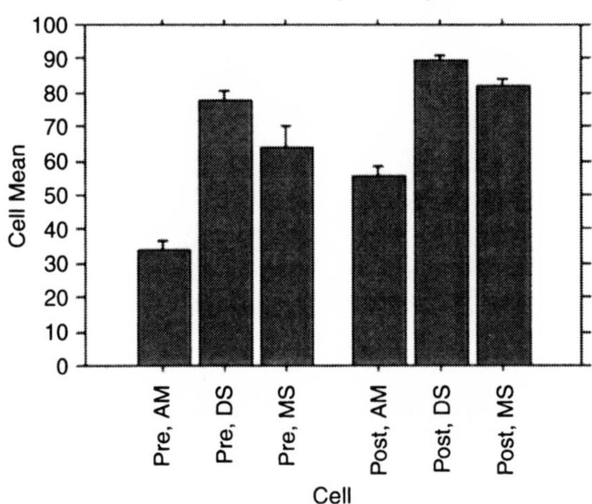

Figure 7.1 Average correctness, all naïve subjects

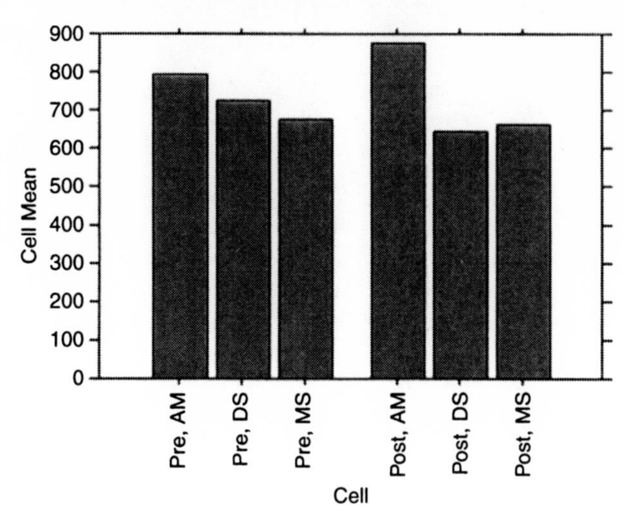

Figure 7.2 Average latency, all naïve subjects

speech has the highest average identification score, at 89%. The difference in subject responses between pre-implant and post-implant tokens is significant (p=.003), as is the difference in subject responses to each speaker (p<.0001).

The average latency results indicate that the speaker with the lowest identification scores (AM) has the highest latencies—the harder it is for a listener to identify speech, the longer it takes. ANOVA did not produce any significant results for this data.

7.3.2 Data exclusions

Looking at the individual subject responses, two criteria for subject exclusion become clear. First, two subjects have several average latencies over 1500 msec; they are clearly not responding "as quickly" as possible (as per subject instructions). Excluding their responses should give a more accurate picture of average response time. Second, one subject responded with "post-implant" for nearly every token from speakers MS and AM. This response bias lowers the pre-implant correctness scores and raises the post-implant correctness scores for these two speakers; excluding that subject's responses should give a more accurate picture of average correctness.

These exclusions do not have a strong effect on average correctness; most scores adjust by one or two points. The significance of subject responses remains the same (p<0001) and the significance of pre-implant and post-implant responses is slightly lower (p=.0057).

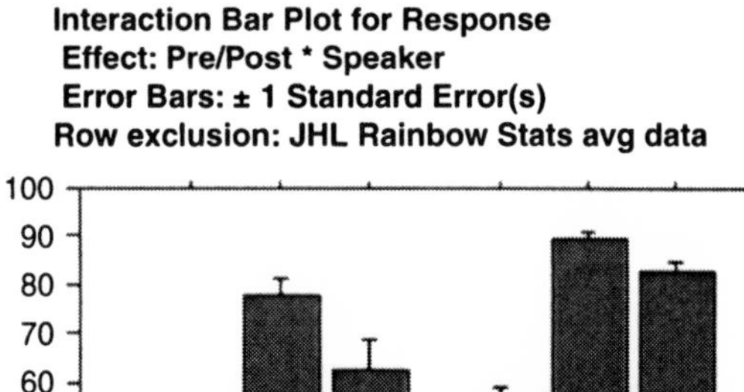

Figure 7.3 Average correctness, adjusted for high latencies and response bias

These exclusions have a strong effect on average latency; latency is approximately 100 msec lower for each category. Speaker AM's pre-implant tokens no longer have the highest response time. The interaction of pre-implant/post-implant and speaker also becomes significant ($p=.0428$).

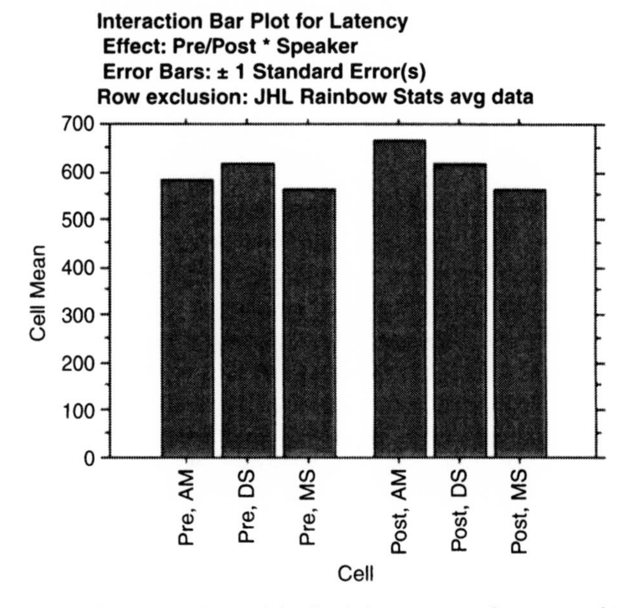

Figure 7.4 Average latency, adjusted for high latencies and response bias

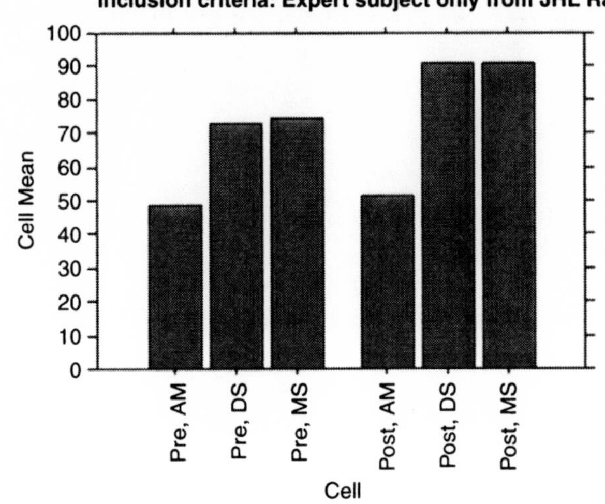

Figure 7.5 Average correctness, expert listener.

7.3.3 Expert listener

The expert listener is fairly accurate in her responses to speakers DS and MS; 73% correct in identifying their pre-implant tokens and 91% correct in identifying their post-implant tokens. On the other hand, her responses to AM's tokens are nearly at chance: 48% correct for pre-implant, 51% correct for post-implant.

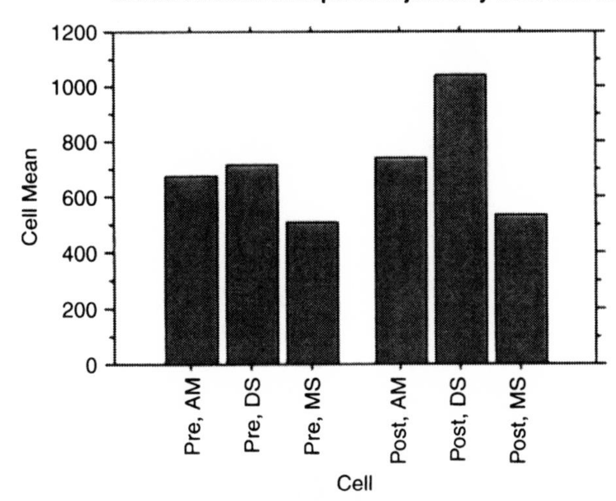

Figure 7.6 Average latency, expert listener

The expert listener's lowest average latencies are for her responses to speaker MS, and her pre-implant and post-implant latencies are nearly the same. This may indicate that classification of MS's tokens was a relatively easy task. On the other hand, her average latency for DS post-implant tokens is over 1000, perhaps corresponding to a more difficult task. Her average latency for speaker AM is around the same for pre-implant and post-implant tokens.

7.4 POSSIBLE ACOUSTIC EXPLANATIONS FOR PERCEPTION RESULTS

Overall, speaker DS's tokens have the highest pre-implant and post-implant correctness scores; this may result from several factors. Her change in MSD is significant for pre-implant vs. six months; at six months she tends to produce higher F0s at the start of Rainbow Passage phrases and lower F0s at the end of Rainbow Passage phrases, as well as change which syllable has the highest F0; and perhaps most crucially, the reduction in duration between pre-implant and post-implant is significant for all Rainbow Passage phrases.

AM's tokens have the lowest pre-implant and post-implant correctness scores. Looking at his acoustic data, the difference between his pre-implant and six months MSDs is not significant. The pattern of change and return in his Rainbow Passage F0 graphs is also telling; his intonation for pre-implant and for six months is very similar, at least for the first part of most phrases. He does demonstrate a significant difference in overall Rainbow Passage duration comparing pre-implant and six months, but this change is only significant for four of the individual phrases (5, 6, 9, 11). These are the likely figures that contribute to the low correctness scores for his tokens.

The results for MS's tokens are less straightforward, at least when looking at MSD, F0, and duration. Her tokens are perceived at 66% for pre-implant, and 81% for post-implant. Yet her MSDs do not significantly change between pre-implant and post-implant, nor does her overall Rainbow Passage duration. Listeners, however, may be picking up on the pattern of lower F0s post-implant, as well as changes in which syllable has the highest average frequency. Other factors not measured may also play a role, such as nasality.

7.5 SUMMARY

Overall, DS's speech is the easiest for naïve listeners to categorize, MS's speech is slightly more difficult, while AM's is the hardest. The expert listener's responses follow the same pattern. These results can be directly related to acoustic analysis of each speaker's Rainbow Passage MSD, F0 and duration.

8 Cochlear implants on U.S. television

8.1 INTRODUCTION

Cochlear implants were featured in both documentary and dramatic presentations on U.S. television between 1998 and 2002. The 2000-2002 television seasons saw three different documentary presentations of cochlear implants and surrounding issues. *Sound and Fury*, which aired on PBS in January of 2002 (as well as at film festivals in 2000 and 2001), presents the complex story of the Artinian family: one brother, Peter, is Deaf,[1] with a Deaf wife and three Deaf children; the other brother, Chris, is hearing, with a hearing wife from a Deaf family, and twin infant sons, one of whom is deaf. The hearing brother want his son to get a cochlear implant, and the Deaf brother's 5-year-old daughter also becomes interested in getting a cochlear implant. The brothers and their families clash over what is the right thing to do for each child. The fifth episode of the 6-episode documentary series *Hopkins 24/7*, which aired September 21, 2000 on ABC, told the story of three cochlear implant recipients: a man who lost his hearing in his 50s, a woman in her 30s who has been deaf since birth, and a 5-year-old child who started losing his hearing at age 2. The "Spare Parts" segment of *Superhuman Body*, which aired March 27, 2001 on The Learning Channel, depicts the experience of Kyle, age 2, getting a cochlear implant.

The 1998–1999 television season featured two series with an extended cochlear implant-related storyline. On the 98–99 season of the NBC medical drama *ER*, Dr. Peter Benton's son Reese (a toddler) is diagnosed as having a severe hearing impairment, and his father seriously contemplates a cochlear implant for his son. During the last week of May and first weeks of June, 1999, the CBS soap opera *Guiding Light* dedicated a major plotline to the story of regular character Abby Bauer, a deaf adult who speaks and lipreads, getting a cochlear implant and her first experiences once it is turned on. In January of 2001, *Gideon's Crossing*, an ABC medical drama, told the story of a 14-year old deaf girl who wanted to get a cochlear implant in spite of the objections of her Deaf parents. Each of these shows handles the topic of cochlear implants very differently; much of this difference has to do with the greater controversies surrounding the implantation of young deaf children, but it is also clear that the writers of each show had different agendas and different priorities.

This chapter specifically analyzes these documentaries and dramas for what kind of information about cochlear implants is being transmitted, sources and accuracy of information, and what kind of communicative expectations might arise from viewing these television series.

8.2 DOCUMENTARIES

8.2.1 *Sound and Fury*

> "I made the film...as a plea for acceptance on both sides of the fence, so that parents of deaf children can make their own decision about what to do. Every parent has the right to do that, every parent has the obligation to do that, and every situation is different."
>
> Josh Aronson (director), Chicago International Film Festival panel discussion, October 7, 2000.

Sound and Fury, directed by Josh Aronson, focuses on several months in the life of the extended Artinian family, who live on Long Island in New York. Peter Artinian is Deaf, his wife Nita is Deaf, their three children Heather (4 1/2), Tim (2 1/2), and CJ (infant) are Deaf. Peter's brother Chris and his wife Mari, who are both hearing, have twin infant sons, Peter (deaf) and Christopher (hearing). Mari's parents (Nancy and Michael Mancini) are both Deaf; Peter and Chris's parents (Peter (Sr.) and Marianne) are both hearing. Chris and Mari are looking into getting a cochlear implant for their son Peter; Heather has also expressed interest in getting a cochlear implant.

Chris and Mari visit audiologists, a cochlear implant surgeon, a special preschool class with several implanted children; they have heated discussions with Mari's parents, who see Mari as forcing her child not to be Deaf. Nita becomes interested in getting a cochlear implant herself, but feels discouraged when the audiologist tells her it probably won't work well for her, as a prelingually deafened adult.

Peter, Nita, and Heather travel to meet two different families: the Hajdamachas, a Deaf family whose ten-year-old daughter Nancy has a cochlear implant, and the Vestrichs, a hearing family whose five-year-old daughter Shelby has a cochlear implant (since the age of 2 1/2). Nancy Hajdamacha likes her cochlear implant—she can use the phone and understand her hearing friends. But she primarily uses ASL to communicate, and goes to school with other Deaf students. When she speaks, she is difficult for the audience to understand. The Artinians enjoy meeting with the Hajdamachas, who understand their cultural perspective.

Shelby Vestrich "views herself as a hearing person" (as described by her father)—she only communicates using speech, and she has no idea what it means when the Artinians use ASL to communicate. Shelby's parents can't understand why the Artinians might decide not to implant Heather; the Artinians can't understand why the Vestrichs would not expose Shelby to ASL. Nita's comment: "It was a mistake to meet them." After these visits, Peter and Nita decide against a cochlear implant for Heather; instead, they will work to get her a good education. We see Nita tell Heather why they decided that Heather shouldn't get an implant, and Heather responds that she doesn't care that she's not getting one.

At a family barbecue, Marianne Artinian is told of Peter and Nita's decision, and she objects strongly—she thinks Heather should be implanted, that it would make her life easier, that Peter and Nita don't know what's best for their child. Peter replies: "I didn't know you didn't accept Deafness until now." We next see the extended Artinian family at the Mill Neck Manor School for the Deaf harvest festival, where they interact with other members of the Deaf community, who discuss cochlear implants and rousingly agree when one man asks, "Do we think the implant can kill Deaf culture, yes or no?" At the elder Artinian's home, Peter and his father argue; his father believes that deafness is a handicap and should be fixed, and accuses Peter of being an abusing parent:

> If I didn't know you, I'd say you were an abusing parent, because you have an opportunity to take a handicap and correct it. You are preventing a cure for deafness to take place because you are so involved with the Deaf world that you really want your children to continue to be Deaf and if they function in the hearing world, okay, but they've got to be Deaf first. That's wrong.

His father tries to draw a parallel with 'cripples,' which Peter disagrees with strongly: "I don't think Deafness is a handicap. I just can't hear."

Peter and Nita decide to drive their family down to Fredericksburg, Maryland, to see the Maryland School for the Deaf. They are amazed by how aware of Deaf culture the city is—a clerk at the grocery store signs an aisle number, and the head of the school tells them that many doctors and other area professionals sign. Peter decides to move his family to Maryland: "I come here to the Deaf world and I feel safe." When Peter announces their upcoming move to his parents, Marianne accuses him of limiting his children and moving just to escape the issue of cochlear implants. We next see Peter and his family moving into their new house, in Maryland.

Chris and Mari are at the hospital for baby Peter's implant surgery, along with both sets of grandparents, Peter (Sr) and Marianne Artinian, and Nancy and Michael Mancini. Scenes of the surgery are interspersed with family discussion in the waiting room. Marianne tries to comfort the Mancinis: "People will never forget what Deafness was, because it'll be history." But Nancy is afraid that baby Peter will not be able to understand her, will make fun of her. One month later, the cochlear implant activation session goes very well, and we see Chris and Mari using sound to interact with baby Peter. The closing credits run over footage of the twin's first birthday party.

I note that there are some differences between the 80 minute theatrical cut of *Sound and Fury*, and the 55 minute PBS cut, primarily involving shorter edits of the family arguments and different closing credit footage.

Overall, *Sound and Fury* accurately portrays the controversy surrounding cochlear implants in deaf children, and portrays both sides in a balanced manner. The medical information presented about cochlear implants is accurate; the two children with CIs, however, really represent two ends of a wide spectrum of hearing/speech ability in cochlear implant users. Shelby is an exceptional cochlear implant user—we even see her singing in the classroom and at home, and in mainstream classes when she starts kindergarten. Nancy, on the other hand, does not seem to have undergone extensive speech/hearing training, and is primarily living in a Deaf world. Most children with cochlear implants would fall somewhere between these two extremes, particularly in the understandability of their speech.

8.2.2 *Hopkins 24/7*

Hopkins 24/7, which aired on ABC in August and September of 2000, is described by the producers as a documentary series that "demystifies modern medicine, showing doctors in rare moments of candor and vulnerability as they share their most closely held thoughts on work, patients and life itself" (*Hopkins 24/7*). The fifth episode, which aired September 21, 2000, presented the story of three cochlear implant recipients: Joe, 61, who was deaf in one ear since birth and has recently lost the hearing in his other ear; Faith, a native signer in her mid-30s who has been deaf since birth; and Cody, a 5-year-old whose progressive hearing loss was diagnosed at age 2. Throughout the episode, the stories of these three patients are interwoven with those of patients in the Eating Disorders Program and that of Dr. Tamara Dildy, a resident in the ER, and take up approximately ten minutes total of screen time.

The cochlear implant segment opens eight minutes into the episode, with footage of Joe meeting with the cochlear implant clinic audiologist. Dr. John Niparko, the head of the Johns Hopkins cochlear implant program, shows the audience an implant and is quoted as saying that "96% of his patients end up hearing successfully." A voiceover asks if a cochlear implant could be successful for someone who has had profound hearing loss since birth, and we meet Faith, who "has always yearned to be part of the speaking world". Cody is then introduced, and his mother discusses his behavior problems and her desire for her son to hear something, even if it's not very much. The next segment opens with Cody, Faith, and Joe each being wheeled in for the surgery. The voiceover:

> "On three different days, Cody, Joe, and Faith are all going in for the same cochlear implant surgery, a two hour procedure. After recovery from the operation will come months of practice, adapting to their new sense of hearing. And how these three patients adapt to hearing with the implant will vary as widely as their three different backgrounds."

The audience is then reminded that there will be five weeks of recuperation time. The next segment consists of the cochlear implant activation sessions for all three patients. Each reacts differently: Joe makes faces as he talks and hears himself, looks as though the experience is painful, and asks to turn it off; Cody puts his hands over his ears; Faith describes the input as sounding like a machine, like noise. The next segment takes place 24 hours after the cochlear implant has been turned on. Joe can't understand speech, and is impatient—he wants the implant to work right away. Faith loves it. Cody's mom describes how he said "off" when a loud radio was playing at home. They then show footage of all three patients working with the speech and hearing therapist: "A long process has begun. It will take nearly two years, but even the once profoundly deaf can expect to achieve 60% comprehension." The end of the episode presents a screen of follow-up text for each of the stories focused on in the episode: "Nearly a year after receiving their cochlear implants, Faith, Joe, and Cody are all making progress. They are learning to understand and produce speech...each at their own pace."

Despite the fairly small amount of time dedicated to a complex issue, the depiction of the cochlear implant experience in this documentary is fairly realistic, at least in terms of the three patients presented. The audiologist during the first segment tells Joe that he may be frustrated when the implant is first activated; this is certainly the case in his experience. There really is no discussion about controversial aspects of cochlear implants in children and the opinions of the Deaf community. The descriptions of cochlear implant success are realistic, but the prospect of failures is not mentioned. It would have been informative to have some footage for one year after the implant activation, rather than just the short follow-up text.

8.2.3 *Superhuman Bodies: Spare Parts*

This Learning Channel/BBC documentary series focuses on "the human body's enormous capacity for self-repair and reinvention and the amazing steps taken daily by modern medicine" (*Superhuman Bodies*). The second hour of the series, *Spare Parts*, discusses cochlear implants, a high-tech prosthetic arm, hand transplantation, heart transplants, and cross-species transplants. *Spare Parts* opens with the story of Kyle Duxbury, a profoundly deaf two-year-old who loves drawing and playing. His mother discusses how her son's deafness was diagnosed (an auditory brainstem response test, or ABR). Close-up footage of the ear and electron microscope pictures of the cochlea and of hair cells accompanies the narrator's description of how hearing works. The narrator describes the cochlear implant as a "bionic replacement", and holds up both an actual cochlea and a cochlear implant for comparison. There is extensive footage of Kyle's actual surgery. We resume a month later with Kyle's visit to the audiologist to have the implant activated. The implant is turned on, and he looks confused and upset. Six months later, we see Kyle sitting on his mother's lap and vocalizing; he is learning to speak. The segment ends with Kyle waving and saying "bye bye" to the camera.

Again, this is a fairly straightforward depiction of the cochlear implant process, at least in terms of surgery and implant activation. Controversial aspects of implantation are not mentioned in the segment.

8.2.4 Discussion

The primary focus of the medical documentaries *Hopkins 24/7* and *Superhuman Body: Spare Parts* is the presentation of factual information about the cochlear implant experience, and *Hopkins 24/7* depicts a wide variety of implantees. *Sound and Fury*, on the other hand, is primarily focused on presenting both sides of the controversy surrounding the Deaf community and cochlear implants in children. All of these documentaries present accurate medical information about cochlear implants, such as candidacy information, what cochlear implant surgery involves, and what happens when the cochlear implant is activated.

Hopkins 24/7 and *Superhuman Body* do not depict any experienced cochlear implantee, so they do not transmit any significant communicative expectations. In *Sound and Fury*, we see two extremes of experienced child CI users: Nancy Hajdamacha, a Deaf child from a Deaf family, whose speech is very difficult to understand; and Shelby Vestrich, who was implanted at 2 1/2, and who at age 5 has extremely clear, "normal" speech. Shelby is an exceptional CI user; speech abilities in implanted children generally fall between these two extremes. Watching *Sound and Fury* might result in unrealistic expectations that any implanted child from a hearing family will have Shelby's speaking and listening skills.

8.3 TELEVISION DRAMAS

8.3.1 Soap opera: *Guiding Light*

From May 28, 1999 to June 16, 1999 a major plotline in the CBS soap opera *Guiding Light* concerned the character Abby Bauer getting a cochlear implant and her experiences after it is turned on. Abby Bauer is played by Amy Ecklund, a deaf actress who had cochlear implant surgery approximately four months before her character did. In the May 27 episode, Abby is in the hospital, soon to have surgery. Her friend Priscilla, who is deaf, visits and tries to convince Abby not to have the implant. Abby talks (using sign and voice simultaneously) with Priscilla again, and tells her that even with the implant, she'll still fight against discrimination, and whether or not the implant works, she'll still be Priscilla's friend. Priscilla wishes her luck, and leaves. Abby and Rick discuss the surgery and their commitment to one another no matter what the results. The nurse comes in to tell them it's time for the surgery, and Abby is wheeled off.

The May 30 episode opens with Rick pacing in Abby's room, waiting for her to come out of surgery. He confides to his sister Michelle, "There's no guarantee it's going to work." Rick tells Michelle that the doctor told him and Abby that they'll need to see a psychiatrist, because there will be so much change in their relationship from having the surgery. Abby is wheeled back to her room after surgery, and wakes up briefly. The nurse tells him that the procedure went as planned, and Michelle is anxious to find out how soon Abby will be able to hear; Rick tells her that it's different for everyone.

Abby wakes up again at the start of the May 31 episode. Rick tells Abby that the surgery went well, and asks her if she can understand him. Abby is groggy and complains about pain. The surgeon comes in to check on her and tells them that everything went perfectly.

Early in the June 1 episode, Rita and her husband Josh are visiting Abby in the hospital. Rick talks with Josh in the hallway, telling him that given Abby's insistence that she's fine the way she is, "I'm don't think my wife's ready to let herself hear." Rick thinks the possibility of hearing again scares Abby—maybe he wanted the surgery more than she did. Abby wakes up, and asks her husband if something is wrong. He reminds her that she'll have to heal before they turn on the implant; she reminds him that it was her own decision to have the implant, though she'd never have had the courage without him. Rick leaves to ask the doctor if Abby can go home. Abby wonders if it's too soon for her to go home, and then she falls asleep. In her dream, the implant doesn't work; she can't hear people, or music, or the telephone, or the TV, even though she knows they're making noise. She wakes up from this nightmare looking shaken.

The June 3 episode opens with Abby and Rick at home. They get a phone call from the doctor, who wants them to come in that night to activate the implant. Abby is scared and apprehensive. They go to the doctor's office, where the doctor checks to see if the incision has healed and shows them the sound processing unit. He puts the earpiece behind Abby's ear and the transmitter on her scalp, then plugs the cord into a computer for a systems check and program check. Abby alternates between saying she's ready and she's not ready. The doctor tells her: "The moment the speech processor is on, you'll be able to hear." When the scene resumes, Abby says she's hearing beeps; the doctor tells her that's the computer setting the programming. The doctor unhooks her from the computer, attaches the cord to her speech processor and has her turn the processor on. She starts when she hears their voices, and exclaims; she then realizes she's hearing herself: "That's me!"

The start of the June 7 episode is a reprise of the last scene from June 3. The scene resumes with Abby commenting on the sound of pulling tissues out of a box: "Paper is very loud!" She asks her husband to talk, so that she can listen to his voice, and keeps pulling the tissues out of the box to hear the sound they make. The doctor tells them to take it slowly. Abby and Rick go home, and Abby finds everything incredibly noisy, asking what specific sounds are coming from. She sits down on the bed, Rick sits down behind her and tells her he loves her—and she is amazed that she understands his speech. She starts to play with her own voice and with making noises with things around their room. Later, she asks what the noise is coming from outside, and Rick tells her that it's crickets. When we return to the scene, suddenly Abby is unable to hear. Upset and scared, she tells Rick the implant stopped working.

The June 8 episode opens with Abby and Rick at home in bed. She is upset, saying "Maybe this wasn't meant to be." Rick tells her not to give up, and she tells him it's her decision; she doesn't know what she wants, or whether she wants to go back the noisy world the implant gave her. Rick calls the doctor to tell him what happened. The doctor visits them at home and tells them this might be a rare microphone problem. Abby doesn't want to deal with it right now, so the doctor leaves a replacement microphone and processor for Rick to use when she's ready. Later, she decides that she wants to try it again, Rick hooks up the microphone and processor, and Abby can hear again.

In the June 10, 13, 14, and 15 episodes, Abby's new hearing plays a significant role in a multiple character plotline. June 10 sees Abby meeting with Rick's sister Michelle and her husband Danny at a restaurant. Rick warns Michelle not to shriek, and when Abby comes in, she comments on their voices. Abby and Rick join Danny and Michelle at a special Mass. At the end of the episode, the camera reveals a bomb beneath one of the pews. During the service in the June 13 episode, Abby starts making faces—she hears something ticking. She asks Rick what it is, and he thinks it's just some noise she isn't used to. But she keeps hearing it, so she yells out, looks for the noise, and finds the bomb. After a cliffhanger ending, the bomb is disarmed in the June 14 episode. During the June 15 episode, Rick tells Abby: "If it wasn't for you, that bomb would've killed all of us." After this dramatic event, Abby's plotlines are integrated into the soap opera as a whole, and no longer focus on her cochlear implant.

Overall, this storyline is an accurate portrayal of an adult getting a cochlear implant, though it takes place in the sped-up world of soap opera time. The surgery itself is well-described, and the authors are careful to have the characters discuss their expectations realistically. The primary victim of soap opera time is the sequence of turning the implant on; it's only a couple of days since the surgery, whereas in real life, implantees wait four to eight weeks for the incision to heal before the implant is turned on. It's clear that the portrayal of Abby's first experiences with sound are based heavily on Ecklund's own; in a *People* magazine article written by Amy Ecklund, she emphasizes how loud she found sounds and how she played with sound, and these elements were particularly present in the June 7, 1999 episode. The speed at which Abby was able to understand voices is also based on Ecklund's experiences, but is not true for many adults who get cochlear implants. Beverly Biderman, in her biography *Wired for Sound: A Journey Into Hearing*, tells of how voices sounded when her implant was first activated:

> When he [her audiologist] had tested each electrode, they had sounded like whistles or horns. Now, with them all turned on, I hear a cacophony of horns and whistles going off together. David's voice, when I realize it is a voice I am hearing, does not sound like Donald Duck or Mickey Mouse as David had warned it might. It sounds like Harpo Marx working his horns and whistles (6).

A December 19, 1998 AP article by Kathy Gurchiek describes 59 year old Catherine Clardy's experience after her implant is turned on:

> "I was sitting there...and then all of a sudden it sounded like a bowl of popcorn popping...it was the sounds and the words and such. They were just moving around and they didn't make any sense to me for three to five minutes. It was unbelievable."

The soap opera's portrayal of the speed at which Abby understands speech could result in some unrealistic expectations when people encounter someone with a recent cochlear implant. Also, in listening to Ecklund's speech, there is a definite change between the May 28 episode (most likely filmed with her not wearing the speech processor, which would affect how her speech sounds) and the June 16 episode; this makes it seem as though having a cochlear implant activated can have an effect on speech in a matter of weeks. The audience is not told that the actress received her implant in January 1999, and has had it activated since approximately March 1999, so this change in speech is reflecting three months of experience rather than three weeks.

8.3.2 Medical drama: *ER*

The Reese deafness storyline began early in the fifth season of *ER* (1998-1999). In the second episode of the season, Dr. Peter Benton brings in his son Reese (played by Matthew Watkins, an 18 month old deaf actor) to have his hearing tested, at the recommendation of his son's physician. The visit to the audiologist reveals that Reese has a severe hearing loss, particularly in the high frequencies. The audiologist gives him the news that Reese will be able to be helped by digital hearing aids, which will give him the input to develop language skills. Benton objects to the audiologist's conclusions, and wants his son to be seen by a doctor.

In the next episode (503), Benton meets with a doctor, who confirms the audiologist's findings. Benton brings up cochlear implants, but the doctor tells him that his son is not a candidate for an implant (according to FDA regulations) and that he can be helped by less invasive methods: digital hearing aids and intensive speech therapy. When Benton asks about sign language, the doctor replies: "Why limit him to signing, when we can teach him to hear and speak?" The following episode finds Benton and Reese at the audiologist, who is teaching Benton how to give his son verbal stimuli. Benton seems to find the process frustrating, and Reese sits on his lap, looking confused.

In episode 506, Benton, carrying Reese, runs into Benton's colleague Kerry Weaver, who notices Reese's hearing aids, and asks Benton which language approach he was planning to use—speech reading, or sign? Benton replies that he is looking for an aggressive surgeon who will do a cochlear implant, and Weaver suggests he talk to Lisa Parks at University of Chicago, an "expert in the field." Benton meets with Dr. Parks[2], and is surprised that she is a family practice doctor who is Deaf, and uses an interpretive assistant. She agrees to talk with him about cochlear implants, and other possible options for his son. During their conversation, they reach an impasse:

Dr. Parks: Have you ever thought being Deaf might not be so terrible? [signed, voiced by interpreter]
Benton: So you're saying there's no problems.
Dr. Parks: You can never know what it's like being Deaf, like I can never know what it's like being black.
Benton: I don't look at me being black as a burden. That's who I am.
Dr. Parks: And I don't look at my Deafness as being a burden. That's who I am. If you turn away from your son's Deafness, you will be robbing him of his identity.

Benton sees the cochlear implant as a way to fix his son's "medical problem," while Dr. Parks tries to convince him that Deafness is something more than that.

In the next episode (507), Benton meets with David Kotlowitz, a cochlear implant surgeon, and they discuss cochlear implants:

Kotlowitz: The cochlear implant is nothing short of a miracle, and the technology keeps improving.
Benton: Yeah, that's what I've read.

Kotlowitz: Don't read, talk to people who have it, talk to the parents of—in fact, I can get you in touch with people who have faced exactly what you're facing now.

When Benton mentions that he's "heard an earful" from opponents of cochlear implants, Kotlowitz calls them "fanatics":

> Someone actually used the word "genocidal", that I was trying to eradicate deaf culture. Deafness is not some alternative lifestyle, and the people who think it is, they're beyond reasoning with.

He proceeds to accuse the signing deaf of laziness. Benton seems surprised by the entire diatribe, responding with "Well. Tell me how you really feel." Kotlowitz ends the meeting by asking Benton to bring in his son, to see if he's a candidate. At the end of the episode, Benton is watching Reese play with a toy phone, and tells his sister "...I feel like I'm in the middle of a battlefield. The more information I have, the less I know what to do."

In episode 509, Kotlowitz invites Benton to observe cochlear implant surgery on a three-year-old, Patrick. He meets Patrick's parents, the Shinoharas, and talks to them about their decision, which seems well thought out: "We know this isn't a cure." Benton and Kotlowitz walk into the operating room, and Benton watches Patrick being prepared for surgery. Kotlowitz tells Benton he can fit Reese into his surgical schedule the following Tuesday; Benton voices his concerns, including Reese's age and amount of residual hearing. Kotlowitz dismisses them and tells his nurse to add Reese to his schedule. Kotlowitz describes the surgery in a fair amount of detail, pointing out where he will make the incision and how he will proceed, and the camera focuses on Patrick as Kotlowitz makes the initial incision. The camera pans to Benton, who looks disturbed by the procedure. Some hours after the surgery, Benton checks on Patrick in recovery, and then calls Kotlowitz's office and tells him he wants to postpone Reese's operation: "I mean, you know, he's barely within the age range and he has a fair amount of residual hearing and the technology's changing all the time..." Seeing the surgery performed has clearly given Benton second thoughts about his son undergoing a cochlear implant.

In episode 511, Dr. Parks (from episode 506) brings her granddaughter Gwen to the emergency room and asks Benton to look at her. Her granddaughter is hearing, and interprets for her grandmother. Dr. Parks thinks Gwen has appendicitis, but Benton disagrees; Dr. Parks then insists on an ultrasound. Gwen's mother[3], also Deaf, arrives while Benton is performing the ultrasound (discovering that Gwen's appendix is indeed inflamed), and insists that the hospital provide them with an interpreter. He watches the three of them signing, and asks Gwen what a certain sign means—she tells him it's the sign for "Father". He is paged to daycare, where he is told that Reese is being disruptive, and the teacher suggests that Reese should be in a special needs environment. Benton returns to Gwen's room, to tell them that the interpreter has been delayed, but that they need to perform the appendectomy now. Dr. Parks asks to scrub in and watch the surgery, and Benton reluctantly agrees. During the surgery, Dr. Anspaugh talks about Dr. Parks and how difficult her residency must have been, and Benton tells him that it's not appropriate for them to talk about her if she can't understand them. The surgery is successful, Gwen's mom gives an uncomfortable Benton a hug upon receiving the news, and Dr. Parks writes him a note: "Job well done!" At the end of the episode, Benton walks into daycare and starts playing with Reese. He tries to teach the sign for "Father" to Reese, and after a few tries, Reese signs it back. Benton is clearly happy as he plays with his son.

The Reese storyline becomes less emphasized during the rest of the season. In episode 514, Benton meets with a sign language instructor[4]. He seems dismayed that he finds signing difficult when Reese does not, and asks her how often he should meet with her. She suggests 3 times a week, but Benton balks at the price: $150 a week. She suggests he can use the *Signing Naturally* video textbook, but comments that it's better to learn one on one. Episode 516 finds Benton working at a clinic in rural Mississippi. When he has to take a patient to a larger city for treatment, he is able to use that clinic's videoconferencing system to briefly sign with Reese.

Episode 418 opens with Reese and Benton at home. Benton is talking on the phone about trying to get Reese into a special school, but every time he tries to go to the required open house, the surgery schedule changes so that he can't. Reese pulls a sharp knife from the dishwasher; Benton notices, takes it away, and signs "No." Benton decides to head to the hospital, and looks for Reese, who is holding one of his hearing aids over the toilet. He drops it in, and Benton, exasperated, signs "No." Benton's experiences as a parent in this episode can easily be compared with those of any parent of an 18-month-old, but they are compounded by difficulty in communicating.

Episode 520 finds Benton surprised to see Reese's mother's fiancé dropping Reese off at daycare. When Benton says he'll bring Reese home after speech therapy, the fiancé is surprised: "He's still in that? I thought you were going the signing route." Benton responds: "I'm not going any route, I'm just doing what's best for my son." The final episode of the season opens with Reese and Benton at speech therapy. The therapist comments on Reese's progress: "He's doing great—he's responding to vocalizations and vocalizing back more."

Benton starts out in denial about his son's hearing loss, insisting on seeing a medical doctor and still holding out the hope that the audiologist is wrong. When he visits with the doctor, Benton is portrayed as really interested at looking at all the options for his child; he seems very interested in cochlear implants, but he also asks about sign language and digital hearing aids. Dr. Benton's initial meeting with the Deaf family practitioner reveals that at that point, he considers his son's deafness as a medical problem that is temporary and fixable, and is clearly set on trying to fix his son, rather than accepting that his son will always be deaf. The cochlear implant side of the story is portrayed in an interesting way; Kotlowitz may characterize the opponents of cochlear implants as fanatics, but he is clearly as much of a fanatic on the other side of the fence. Benton seems to be walking a tightrope between the two sides, while wanting to do what's best for his son. When Benton witnesses a happy, functional Deaf family (the Parks), he seems to realize that his son learning to sign could allow for a complete and fulfilling life. By the end of the season, Benton has decided to use whatever communication method works best for his son; Reese is progressing well in speech therapy, but he is also being taught signing as a primary means of communication.

That the storyline developed this way is no surprise, considering their primary consultant: Ken Randall, superintendent of the Arizona State School for the Deaf and Blind, a proponent of total communication (both voice and signing) with an emphasis on signing (Mendoza 1998). Mendoza interviews show writer Linda Gase, who describes the story process: they "talked to experts on all sides of the issue...met with educators from a school that advocates the use of implants in children... [and] visited a preschool that uses both sign and speech" (1B). Their goal was to have Dr. Benton follow the same journey.

Examining the storyline as a whole, the writers have managed to portray fairly accurately the journey of a parent who discovers their child is deaf. However, the cochlear implant advocate, as portrayed by Dr. Kotlowitz, seemed over-the-top. This storyline does not transmit any significant communicative expectations in terms of people with cochlear implants, since we never encounter anyone actually using one.

8.3.3 Medical drama: *Gideon's Crossing*

The cochlear implant episode of *Gideon's Crossing*, "Orphans," aired January 29, 2001. The episode starts with a lost-looking teenage girl wandering through the hospital. A security guard finds her, and brings her to Dr. Gideon's office. Her name is Charlotte Warren; she is deaf (communicating primarily with voice and using hearing aids and lipreading to understand speech); and she is interested in getting a cochlear implant. Her parents are opposed to implants: "They don't understand, they're very old-fashioned." She hopes that if Dr. Gideon explains the implants to her parents, they will let her have the surgery. Gideon finds out that her parents don't know that she's there, and he wants to contact them. We next see Charlotte being evaluated by the cochlear implant surgeon, Dr. Fieldstone. Based on audiological tests, she qualifies for an implant, and her hearing aid use makes her a good candidate. Gideon and Fieldstone explain the possible downsides of getting an implant to Charlotte, and she's willing to take her chances with the procedure. Charlotte's parents, accompanied by an interpreter, meet with Gideon and Charlotte in Gideon's office. Her parents are both Deaf;

her mother (Lindsay[5]) communicates primarily through ASL and her father (Howard) uses ASL and voice simultaneously. Her parents are upset that Gideon examined Charlotte without their permission:

Gideon: It was a simple hearing test, similar to the ones I'm sure Charlotte has received in school. The issue I think we're all concerned with here is correcting Charlotte's hearing loss.
Howard: (sign and voice simultaneously) Deafness is not a defect. We're Deaf. Our friends are Deaf. This is our culture, it's who we are.
Lindsay: We've taught Charlotte to take pride in her Deaf identity. If we now change that, what happens to her feelings of self-worth?
Gideon: What will it do to her self-worth to grow up feeling different and isolated?
Howard: You're assuming that it's better to be hearing than Deaf.
Gideon: I'm assuming that it's easier to be hearing than deaf.

Frustrated, Charlotte signs "I'm going to do this with or without you, I don't care what you think," and runs out of the office.

The next scene, Gideon and Fieldstone are walking down a hallway. Fieldstone disparages the Deaf community position on cochlear implants, and suggests taking Charlotte's case to the ethics committee, so that they can take her parents to court. The ethics committee meets, and Gideon argues for the right of Charlotte's parents to determine what is best for their child. The committee vote is 5 to 4 against going to court (Gideon being the swing vote). We next see Charlotte in the hospital cafeteria, eating lunch next to Ollie, a resident. Through their conversation, Ollie learns about Charlotte's situation, and brings it up while talking to Gideon in a hospital lab. Ollie thinks that Gideon should push for Charlotte to have the surgery. Gideon stops by Fieldstone's office, and watches a young girl with a cochlear implant sing "Twinkle Twinkle Little Star."

Gideon changes his vote with the ethics committee, and the case goes to the courts. Lindsay Warren meets with Gideon in his office, and tells Gideon some things he didn't know about Charlotte: she has a much older hearing boyfriend, who is pressuring her to have the operation; she had an STD at age 13; she has a tattoo of a snake on her stomach. Lindsay argues that Charlotte's desire for a cochlear implant is simply another rebellion against her parents. Gideon and Charlotte talk in his office. He confronts her about the boyfriend and the tattoo; she emphasizes how much she wants the implant, then runs from his office in frustration.

Charlotte runs away, and is soon found by the police. Gideon goes to the Warren's house to meet with the family. Gideon mediates between Charlotte and her mother. Charlotte denies that she is only doing this to hurt her parents, and her mother denies that Charlotte has enough maturity to make this sort of decision. Gideon reminds them that Charlotte isn't a woman yet, but she isn't a child either. We discover that Lindsay is afraid of losing Charlotte; once Charlotte reassures her, she agrees to the surgery. The episode ends with Charlotte being wheeled into the operating room, accompanied by her parents, hopeful for the future.

This episode is interesting in the choice of storyline for depicting the controversy over cochlear implants; instead of seeing a parent deal with an infant's deafness, we see Deaf parents dealing with their teenage daughter's need for independence. The construction of this storyline is somewhat odd—given her family background, Charlotte has remarkably good speech and lipreading skills; given Charlotte's age and abilities, why is it necessary for her to get an implant at 14, when she might get the same results at 18? In addition, scientific research has shown that a child with Charlotte's background might not have much success with a cochlear implant: "the prognosis for adolescents and adults with congenital and/or prelingual hearing losses attaining open-set speech recognition is considered to be poor" (Waltzman and Cohen 1999, 84). The sources and background for this episode are also not as clear; the research and experiences behind the *ER* and *Guiding Light* episodes were reported in U.S. newspapers and magazines, and I could find no such information for *Gideon's Crossing*. There may be some influence from the documentary feature *Sound and Fury*, which began to be exhibited at film festivals about four months before the episode aired; the *Gideon's Crossing* scene where a young girl with a cochlear implant understands speech without lipreading and tunefully sings "Twinkle Twinkle Little Star" as Dr. Gideon looks on strongly echoes the footage of Shelby Vestrich singing in *Sound and Fury*. This sort of ability post-implant is not unheard of, but also not average: this is a portrayal of a child who is doing exceptionally well. Viewing this scene might give people unrealistic expectations when they meet a child with a cochlear implant.

8.3.4 Common themes

There are some common themes in all three plotlines. For example, on both *ER* and *Gideon's Crossing* a representative from the Deaf community makes a comparison between Deafness and being African-American:

ER:

Dr. Parks: You can never know what it's like being Deaf, like I can never know what it's like being black.

Benton: I don't look at me being black as a burden. That's who I am.

Dr. Parks: And I don't look at my Deafness as being a burden. That's who I am. If you turn away from your son's Deafness, you will be robbing him of his identity.

Gideon's Crossing:

Gideon: I'm assuming that it's easier to be hearing than deaf.

Lindsay: Would your life be easier if you were white?

Gideon: I don't know.

Lindsay: Would it?
Gideon: It might be easier if I were white.
Lindsay: Would you change your race, if you could, to have an easier life? Would you?
Gideon: That's not the point.

Racial identity is used as a metaphor to illustrate the cultural aspects of Deafness to the viewing audience.

On *Guiding Light*, the interaction between Abby Bauer and her friend Priscilla similarly addresses Deafness as cultural identity. Priscilla is afraid that Abby will turn away from her Deaf friends and tries to convince her not to have the implant, and Abby reassures Priscilla that she will still be her friend, whether or not the implant works.

The way in which cochlear implant surgeons are portrayed varies between the three shows. On *Guiding Light*, the surgeon is portrayed as caring and concerned; when Abby's implant suddenly stops working, he makes a house call to drop off a new processor and reassures her that there's nothing wrong with her implant. He also is always careful to make sure she can clearly see his face, so that it is easier for her to lipread. The implant surgeons on both *ER* and *Gideon's Crossing*, on the other hand, are portrayed as brusque and having little respect for Deaf culture:

ER:
Benton: I've already heard an earful from the opponents.
Kotlowitz: Ah, the fanatics.
Benton: They've got their agendas.
Kotlowitz: The level of passion with which these people will attack you, it's bizarre. Someone actually used the word "genocidal", that I was trying to eradicate Deaf culture. Deafness is not some alternative lifestyle, and the people who think it is, they're beyond reasoning with.
Benton: The thing I'm trying to figure out is, why make an argument against hearing and speaking, you know?
Kotlowitz: Honestly, do you want to know what I think? I think it comes down to laziness. I think a deaf person can function in the hearing world but some of these people, they just don't want to do all the exercises and do all the drills that it takes.
Benton: It's years of hard work.
Kotlowitz: Oh yeah, but they have the energy to attack me and other doctors, so I say the hell with them.

Gideon's Crossing:
Fieldstone: Typical. These Deaf Rights people are zealots.
Gideon: They're very protective of deaf culture.
Fieldstone: Oh please, my ass. Would they be equally protective of their daughter's harelip?

Gideon: Regardless, their position is intractable.
Fieldstone: So what, we'll go around them.
Gideon: What do you mean?
Fieldstone: We'll take it to the ethics committee.
Gideon: Is that really necessary? She can get this herself in four years.
Fieldstone: Oh no, she's already past the prime age. Four more years, she may not be eligible at all.
Gideon: I just don't think making her parents the enemy is the best way to approach this.
Fieldstone: They are making a decision which is not in her best interests. This is not parental prerogative, it's parental neglect.

This portrayal of implant surgeons seems to be a form of dramatic shorthand, where a single scene is used to set up the pro-implant side as fanatical about implants and opposed to Deaf culture. This is a very one-dimensional portrait; most cochlear implant surgeons probably lie between the two extremes of ideal surgeon (*Guiding Light*) and pro-implant, anti-Deaf-culture fanatic (*ER, Gideon's Crossing*).

8.3.5 Discussion

Each of these television dramas transmitted primarily accurate information about cochlear implants, at least in terms of surgical procedure, who might be interested in getting an implant, and the objections that members of Deaf culture have to implanting young children. Less accurate information includes the one-dimensional portrait of cochlear implant surgeons, the sped-up recovery timeline on the soap opera, and the background of Charlotte's character on *Gideon's Crossing*. Each storyline came from a different source: *Guiding Light* was based on the actress' own experiences, *ER* was based on visits to schools which advocate cochlear implants and schools which advocate signing, and the basis of the *Gideon's Crossing* episode is unclear, but may have been influenced by *Sound and Fury*. There were no real communicative expectations that can be drawn from *ER*, but both *Gideon's Crossing* and *Guiding Light* portray somewhat unrealistic implant users; the little singing girl on *Gideon's Crossing* represents an exceptional implantee, and Abby's near-immediate ability to understand speech on *Guiding Light*, though based on the actress' own experiences, is also unusual.

8.4 CONCLUSIONS

Overall, both the documentaries and the dramas presented accurate information about cochlear implants. *Hopkins 24/7* and *Superhuman Body* primarily transmitted medical information, while *Sound and Fury* focused on a balanced presentation of the controversy surrounding cochlear implants in children. *ER* and *Gideon's Crossing* also presented both sides of the controversy, while *Guiding Light* focused on the experiences of a postlingually deaf adult. *Hopkins 24/7*, *Superhuman Body*, and *ER* do not depict any experienced cochlear implant users; *Sound and Fury*, *Guiding Light*, and *Gideon's Crossing* each depict one highly skilled cochlear implant user, and people watching those broadcasts might come away with unrealistic expectations about the ordinary cochlear implant user's speaking and listening abilities.

NOTES

1 In this chapter, I use deaf/deafness to refer to the medical state of having a severe to profound hearing loss, and Deaf/Deafness to refer to the Deaf cultural community, who communicate using American Sign Language (ASL).
2 Dr. Parks is played by Phyllis Frelich, a well-known Deaf actress
3 Gwen's mother is played by Terrylene Frelich, a Deaf actress who is Phyllis Frelich's daughter.
4 The ASL instructor is played by Marlee Matlin, a well-known Deaf actress.
5 The character of Lindsay was played by Marlee Matlin.

9 Discussion

9.1 SUMMARY OF SUBJECT PRODUCTION RESULTS

AM shows an increase in average vowel space post-implant, with a certain degree of change and return. His vowel and vowel word durations are fairly stable, with long vowels tending to have longer durations. His CV syllable duration is also longer post-activation, as are his voiceless VOTs. In terms of fundamental frequency, AM demonstrates a strong pattern of change and return for vowel peak F0, the Rainbow Passage phrases, and the Bev loves Bob sentences. The similarity of his pre-implant and six month speech patterns are confirmed by the hearing listener perception study results.

DS also shows an increase in average vowel space post-implant. Her vowel and vowel word durations decrease, and this change is significant for long vowels; this tendency towards shorter durations is also seen in her CV syllable durations, voiceless VOTs, and Rainbow Passage phrases. DS produces less variant vowel peak F0s post-activation, but her Rainbow Passage data has a higher degree of variation post-activation (as measured using MSD). Hearing listeners are very accurate in distinguishing DS's pre-implant speech from her six month speech.

MS's average vowel space shows a pattern of change and return. Her vowel and vowel word durations are generally stable, with some tendencies towards change and return. Her CV syllables and VOT are longer post-activation, while her Rainbow Passage phrase durations do not change. In terms of fundamental frequency, MS produces more variant vowel peak F0s and Bev loves Bob F0 patterns post-activation, but her Rainbow Passage phrases generally have lower F0s. Hearing listeners are fairly accurate in distinguishing MS's pre-implant speech from her six month speech.

According to these measures, DS's speech changes the most, AM's the least, and MS is somewhere in the middle. That hearing listeners are fairly accurate in identifying MS's speech as pre- or post-implant may result from articulatory changes which were not captured by this study.

The changes in DS's speech are most like those of the subjects in the Perkell studies, while AM and MS seem to follow a different overall pattern.

This difference may result from AM and MS's relatively short period of profound hearing loss, when compared with the much longer period experienced by Perkell's subjects.

9.2 THE ROLE OF FEEDBACK IN SPEECH PRODUCTION: IMPLICATIONS OF CHANGE AND RETURN

Change and return, the tendency of speech measures to change at one and three months and return to pre-implant levels at six months, has interesting implications for the role of feedback in speech production. Change and return is seen in average vowel space for both AM and MS, in vowel and vowel word duration for MS, and in every measure of fundamental frequency for AM. For these subjects, the feedback from the cochlear implant leads to initial changes, but old speaking patterns return by the sixth month of cochlear implant use. These changes do not seem to be related to changes in vowel or consonant perception, as both subjects' perception scores improve from one month to three months to six months.

According to Perkell's two-level model of speech production, vowel space, duration, and fundamental frequency all fall into the 'postural' level, which they theorize responds to auditory feedback to maintain effective communication. According to their findings, the changes in cochlear users' speech are all the result of postural feedback. Why then do these two subjects return to pre-feedback articulatory habits?

It seems most likely that it may simply have taken more time for these two subjects to get used to the input from a cochlear implant, and the measures at one month and three months would then reflect a certain degree of feedback distortion. MS's highly variable intonation post-activation could also be explained by feedback distortion, particularly with her post-implant comments that she was unable to hear her own voice pitch. It could also be the case that for MS and AM, articulatory habits are more deeply ingrained; feedback from the implant leads to initial changes but muscle memory eventually wins out; but this seems unlikely.

I do note that AM and MS both exhibit clear, understandable speech pre-implant, and that DS, the subject with the largest degree of pre-implant distortions, shows normative changes rather than change and return.

To further explore change and return, I plan to explore whether naïve listeners can distinguish AM's pre-implant and one month/pre-implant and three month Rainbow Passage readings. It would also be interesting to look at the speech of these subjects when they have greater than six months of cochlear implant experience, to see if the return to pre-implant articulations holds, and to study the speech of other cochlear implant users with shorter periods of profound deafness.

9.3 REFLECTIONS ON MEASUREMENTS AND MATERIALS

Overall, the acoustic measurements selected for this study proved their worth. Post-activation, the subjects in this study all demonstrate less restricted vowel space; changes in vowel duration; changes in VOT that vary with changes in CV syllable duration; and changes in both peak F0 and average F0. In terms of vowel space, visual examination of the F1 and F2 plane and the vowel triangle metric did not prove as useful for comparison as the average Euclidean vowel space metric. In terms of fundamental frequency measures, both peak F0 and average F0 provided useful information; peak F0, however, does tend to emphasize outliers. In future studies, I plan to utilize average F0 measures. The "Bev loves Bob" task also proved difficult to analyze, particularly considering the difficulty of averaging when subjects respond inconsistently. I do not plan to use that task in future studies.

9.4 COCHLEAR IMPLANTS ON TELEVISION AND COMMUNICATIVE EXPECTATIONS

Generally, the portrayal of cochlear implants in television documentaries and dramas is factually correct and presents the controversy over cochlear implants in children in a balanced manner. The focus on exceptionally skilled cochlear implantees, however, might lead to unrealistic expectations about the ordinary cochlear implant user's speaking and listening abilities.

Of the subjects in this study, DS might be considered an exceptional implantee: like the character of Abby Bauer on *Guiding Light,* DS could understand her husband's speech right away, and DS's speech perception abilities are at a high level even at one month post-activation. It is difficult, however, and possibly pointless, to 'rate' the skills of each of these implant recipients; all three subjects in this study are pleased with their cochlear implant and find that it improves their quality of life, and all three have easily comprehensible speech and good listening comprehension.

The documentary *Sound and Fury* deserves more extensive analysis that was carried out in this dissertation; the issues it raises are truly complex, and in many ways it has become a representative depiction of cochlear implant issues for mainstream America. Since the broadcast of *Sound and Fury,* I have seen the Artinian's decision not to implant their daughter brought up in editorials and columns as an example of parental selfishness and over-the-top disability activism. Instead of "acceptance on both sides of the fence," reactions to the documentary point to an entrenching of positions.

9.5 FUTURE DIRECTIONS

Further study of the change and return phenomenon should provide greater insight into the role of feedback in speech production. In addition to extending the hearing subjects perception study, an examination of the unanalyzed fricative tokens and the spontaneous speech recordings may prove fruitful. There may also be clinical implications of this phenomenon; would the addition of specific speech or hearing therapy counteract the tendency towards return? What kind of therapy would be most effective?

In addition to the analysis of television programs presented here, a study of cochlear implants in print media will expand the picture of what a modern, technologically sophisticated culture expects of cochlear implants and from cochlear implantees.

A Materials recorded

A.1 /HVD/ WORDS

heed, hid, head, had, hod, hud, hawed, hood, who'd.

A.2 /CACC/ WORDS

pat, pack, pap, pad, path; tap, tat, tack, tab, tad; cap, cat, cab, cad, cash; bap, bat, back, bad, bag; dab, dad, dash, dark, dart; gap, gab, gad, gash, gas.

A.3 BEV LOVES BOB SENTENCES

Bev loves Bob. Bev loves Bob. Bev loves Bob. Bev loves Bob. Bev loves Bob? Bev loves Bob? Bev loves Bob? Bev loves Bob?

A.4 THE RAINBOW PASSAGE

When the sunlight strikes raindrops in the air, they act like a prism and form a rainbow. The rainbow is a division of white light into many beautiful colors. These take the shape of a long round arch, with its path high above, and its two ends apparently beyond the horizon. There is, according to legend, a boiling pot of gold at one end. People look, but no one ever finds it. When a man looks for something beyond his reach, his friends say he is looking for the pot of gold at the end of the rainbow (Fairbanks 1960).

B. Recording session orders and interview questions

B.1 PRE-IMPLANT RECORDING SESSION

Recording order:

1. Rainbow Passage
2. CaC(C) words (Presented on cards, with frame sentence visible, word placed in frame sentence, present card set 3 times)
3. Rainbow Passage
4. Vowel words (Presented on cards, with frame sentence visible, word placed in frame sentence, present card set 3 times)
5. Rainbow Passage
6. Bev loves Bob sentences (Presented on cards, present card set 3 times)
7. David Wiesner *Tuesday*, page 6-7 (frogs and birds). "Describe for me what is going on in this picture".
8. Interview questions (pre-implant)

Card presentation orders:

CaC(C) word cards (34 total)

bat	C17	pack	C2
tad	C10	bad	C19
bag	C20	sad	C31
tat	C7	dart	C25
tab	C9	vat	C34
cab	C13	path	C5
bap	C16	cap	C11
gap	C26	pap	C3
cash	C15	gash	C29
dad	C22	tack	C8
dark	C24	gab	C27
gad	C28	tap	C6
dab	C21	cad	C14

shad	C32	cat	C12
gas	C30	pad	C4
dash	C23	back	C18
fat	C33	pat	C1

Vowel word cards (9 total)

heed	V1
had	V4
who'd	V9
head	V3
hood	V8
hod	V5
hid	V2
hawed	V7
hud	V6

Bev loves Bob sentence cards (8 total)

<u>Bev</u> loves Bob?	B6
Bev loves <u>Bob</u>.	B4
Bev loves Bob.	B1
Bev loves <u>Bob</u>?	B8
Bev <u>loves</u> Bob?	B7
<u>Bev</u> loves Bob.	B2
Bev loves Bob?	B5
Bev <u>loves</u> Bob.	B3

B.2 ONE MONTH RECORDING SESSION

Recording order:

1. Rainbow Passage
2. Bev loves Bob sentences (Presented on cards, present card set 3 times)
3. Rainbow Passage
4. CaC(C) words (Presented on cards, with frame sentence visible, word placed in frame sentence, present card set 3 times)
5. Rainbow Passage
6. Vowel words (Presented on cards, with frame sentence visible, word placed in frame sentence, present card set 3 times)
7. David Wiesner *Tuesday*, page 11 (midnight snack). "Describe for me what is going on in this picture."
8. Interview questions (One month post activation)

Card presentation orders:

CaC(C) word cards (34 total)

cap	C11	pad	C4
dark	C24	bag	C20
vat	C34	cat	C12
gad	C28	tat	C7
dart	C25	tab	C9
dab	C21	cad	C14
sad	C31	tap	C6
bad	C19	cab	C13
gas	C30	gab	C27
pack	C2	bap	C16
dash	C23	tack	C8
fat	C33	gap	C26
pat	C1	gash	C29
bat	C17	pap	C3
shad	C32	cash	C15
back	C18	dad	C22
tad	C10	path	C5

Vowel word cards (9 total)

who'd	V9
head	V3
hud	V6
hid	V2
hod	V5
heed	V1
hawed	V7
hood	V8
had	V4

Bev loves Bob sentence cards (8 total)

Bev loves Bob.	B2
Bev loves Bob.	B1
Bev loves Bob?	B7
Bev loves Bob?	B8
Bev loves Bob.	B3
Bev loves Bob?	B6
Bev loves Bob?	B5
Bev loves Bob.	B4

B.3 THREE MONTH RECORDING SESSION

Recording order:

1. Rainbow Passage
2. Vowel words (Presented on cards, with frame sentence visible, word placed in frame sentence, present card set 3 times)
3. Rainbow Passage
4. Bev loves Bob sentences (Presented on cards, present card set 3 times)
5. Rainbow Passage
6. CaC(C) words (Presented on cards, with frame sentence visible, word placed in frame sentence, present card set 3 times)
7. David Wiesner *Tuesday*, pages 16-17 (TV). "Describe for me what is going on in this picture."
8. Interview questions (Three/six months post activation)

Card presentation orders:

CaC(C) word cards (34 total)

gab	C27	shad	C32
path	C5	tap	C6
cab	C13	bag	C20
sad	C31	pack	C2
cad	C14	gash	C29
dad	C22	back	C18
tack	C8	fat	C33
dark	C24	bad	C19
pat	C1	dart	C25
vat	C34	gad	C28
pap	C3	cat	C12
cash	C15	gap	C26
tab	C9	pad	C4
bat	C17	gas	C30
bap	C16	tad	C10
dab	C21	cap	C11
dash	C23	tat	C7

Vowel word cards (9 total)

hid	V2
hud	V6
hood	V8
head	V3
had	V4
hod	V5
who'd	V9
heed	V1
hawed	V7

Bev loves Bob sentence cards (8 total)

Bev loves Bob?	B7
Bev loves Bob.	B2
Bev loves Bob.	B4
Bev loves Bob?	B8
Bev loves Bob?	B6
Bev loves Bob.	B1
Bev loves Bob.	B3
Bev loves Bob?	B5

B.4 SIX MONTH RECORDING SESSION

Recording order:

1. Rainbow Passage
2. Bev loves Bob sentences (Presented on cards, present card set 3 times)
3. Rainbow Passage
4. CaC(C) words (Presented on cards, with frame sentence visible, word placed in frame sentence, present card set 3 times)
5. Rainbow Passage
6. Vowel words (Presented on cards, with frame sentence visible, word placed in frame sentence, present card set 3 times)
7. David Wiesner *Tuesday*, pages 26-27 (investigation). "Describe for me what is going on in this picture".
8. Interview questions (Three/six months post activation)

Card presentation orders:

CaC(C) word cards (34 total)

bat	C17	pack	C2
tad	C10	bad	C19
bag	C20	sad	C31
tat	C7	dart	C25
tab	C9	vat	C34
cab	C13	path	C5
bap	C16	cap	C11
gap	C26	pap	C3
cash	C15	gash	C29
dad	C22	tack	C8
dark	C24	gab	C27
gad	C28	tap	C6
dab	C21	cad	C14
shad	C32	cat	C12
gas	C30	pad	C4
dash	C23	back	C18
fat	C33	pat	C1

Vowel word cards (9 total)

heed	V1
had	V4
who'd	V9
head	V3
hood	V8
hod	V5
hid	V2
hawed	V7
hud	V6

Bev loves Bob sentence cards (8 total)

Bev loves Bob?	B6
Bev loves Bob.	B4
Bev loves Bob.	B1
Bev loves Bob?	B8
Bev loves Bob?	B7
Bev loves Bob.	B2
Bev loves Bob?	B5
Bev loves Bob.	B3

B.5 PRE-IMPLANT INTERVIEW QUESTIONS

1. How old are you?
2. At what age did your hearing loss occur?
3. Was it immediately identified?
4. Do you know the cause of your hearing loss?
5. How is your hearing loss classified (degree of loss)?
6. Has your hearing loss changed over time?
7. What benefits have you received from hearing aids?
8. Have you had any experience with speech therapy?
9. If so, how would you describe your experience?
10. What made you interested in getting a cochlear implant?
11. What are your expectations?
12. How would you describe your experiences with the Deaf cultural community?

B.6 ONE MONTH INTERVIEW QUESTIONS

1. Do you remember your first impressions when the implant was activated?
2. Which speech processor are you using?

3. How much post-implant hearing and/or speech training have you had?
4. What does the signal from the implant "sound" like to you?
5. How does the signal from the implant currently compare to the signal you got from hearing aids?
6. How would you describe your experiences with the Deaf cultural community, after your implant (if applicable)? Is this a change from before you received the implant?
7. What have been the benefits of a cochlear implant, in your experience?
8. Have there been any drawbacks?

B.7 THREE AND SIX MONTH INTERVIEW QUESTIONS

1. Have you had any additional hearing and/or speech training?
2. What does the signal from the implant "sound" like to you? Has this changed?
3. How does the signal from the implant currently compare to the signal you got from hearing aids?
4. How would you describe your current experiences with the Deaf cultural community (if applicable)? Is this a change from before you received the implant?
5. What have been the benefits of a cochlear implant, in your experience?
6. Have there been any drawbacks?

C Rainbow Passage figures

C.1 Subject AM

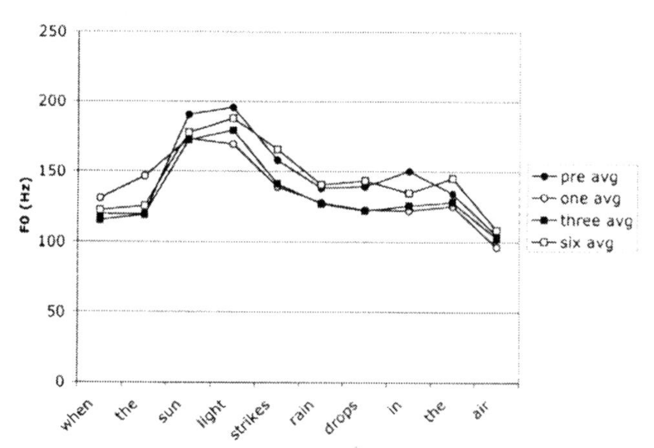

Figure C.1 Subject AM Rainbow Passage, phrase 1: Average F0

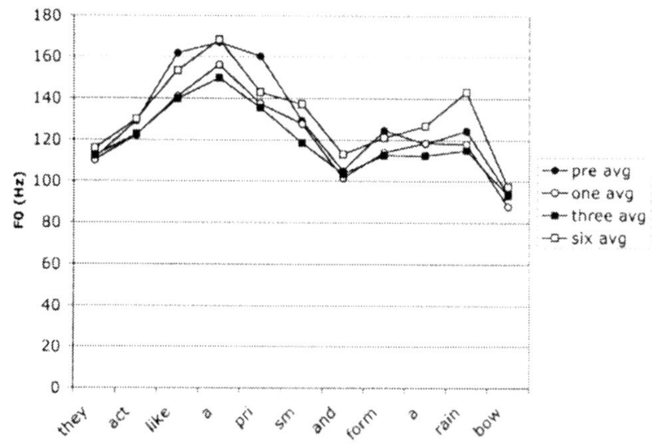

Figure C.2 Subject AM Rainbow Passage, phrase 2: Average F0

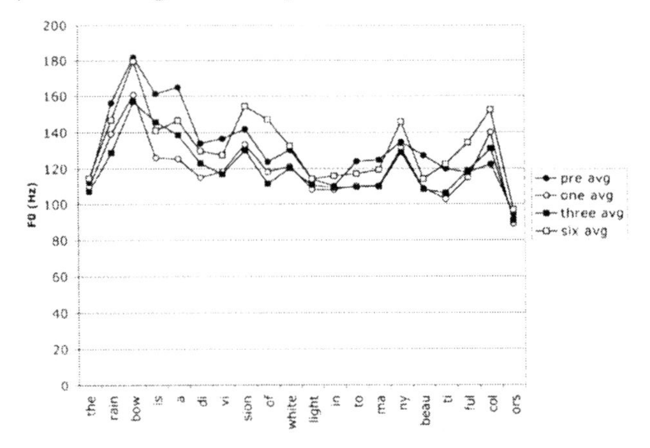

Figure C.3 Subject AM Rainbow Passage, phrase 3: Average F0

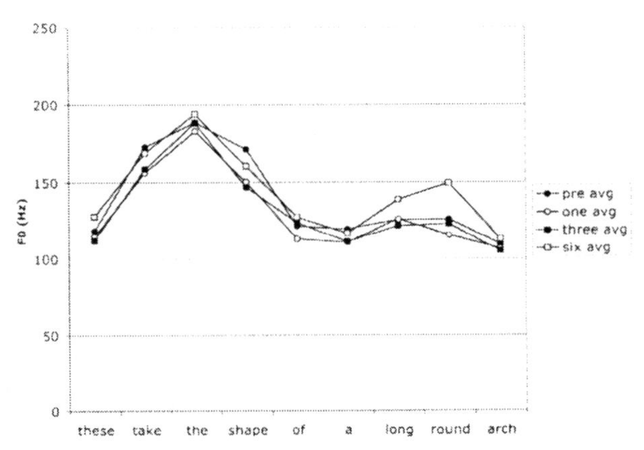

Figure C.4 Subject AM Rainbow Passage, phrase 4: Average F0

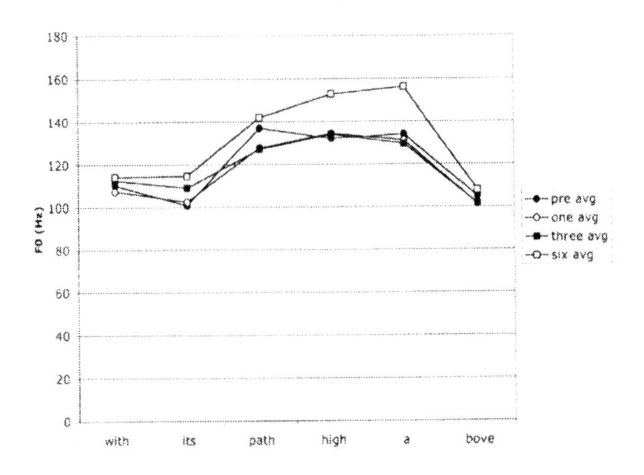

Figure C.5 Subject AM Rainbow Passage, phrase 5: Average F0

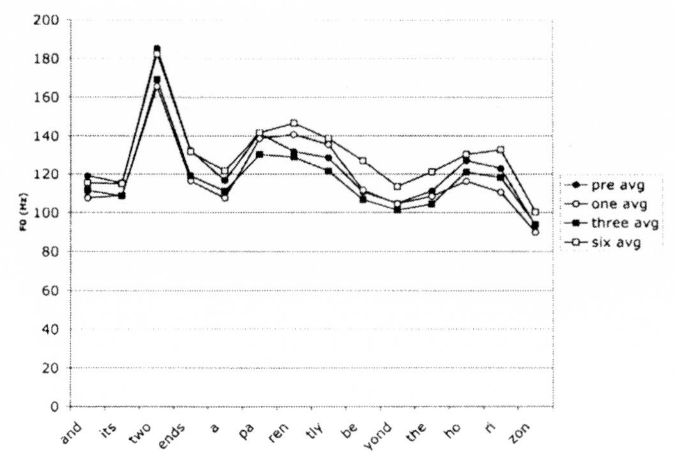

Figure C.6 Subject AM Rainbow Passage, phrase 6: Average F0

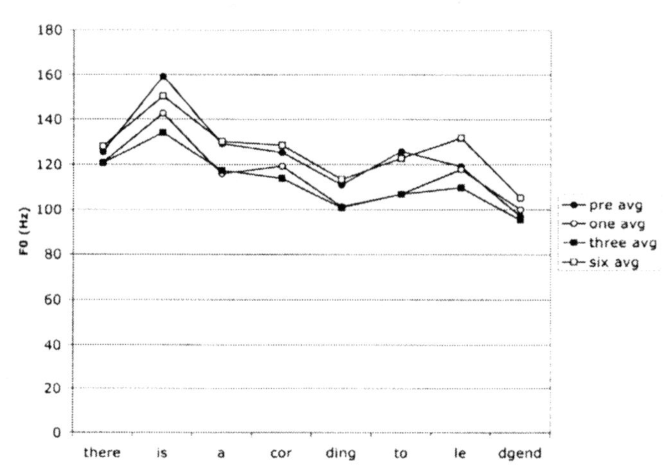

Figure C.7 Subject AM Rainbow Passage, phrase 7: Average F0

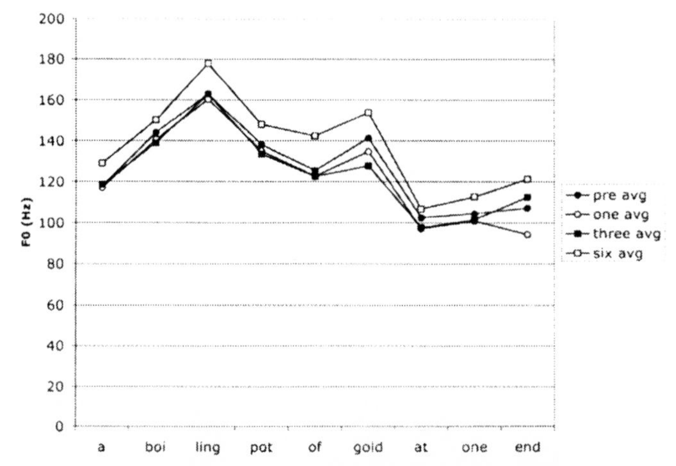

Figure C.8 Subject AM Rainbow Passage, phrase 8: Average F0

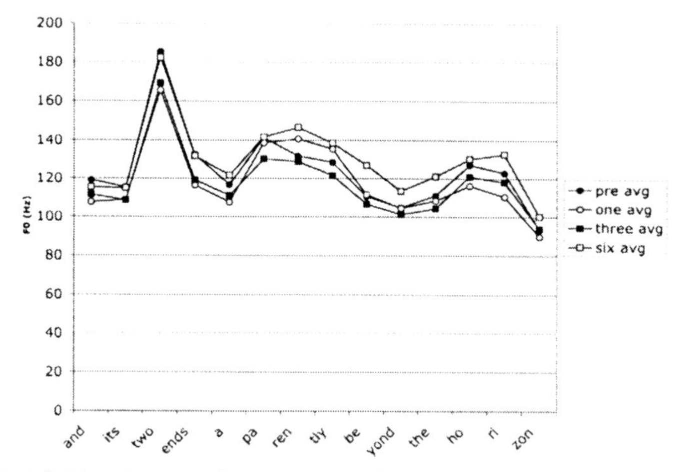

Figure C.6 Subject AM Rainbow Passage, phrase 6: Average F0

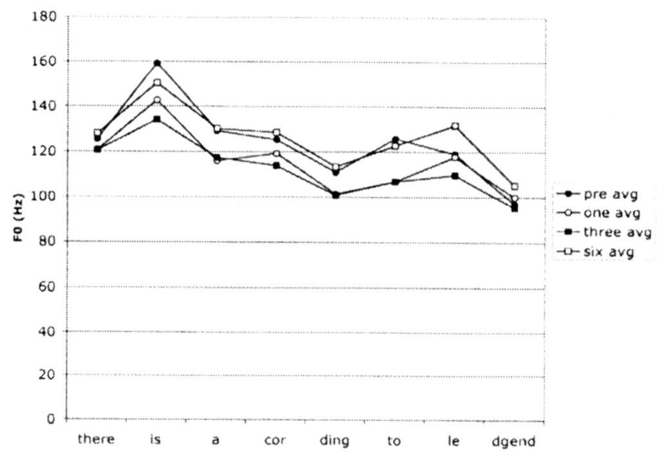

Figure C.7 Subject AM Rainbow Passage, phrase 7: Average F0

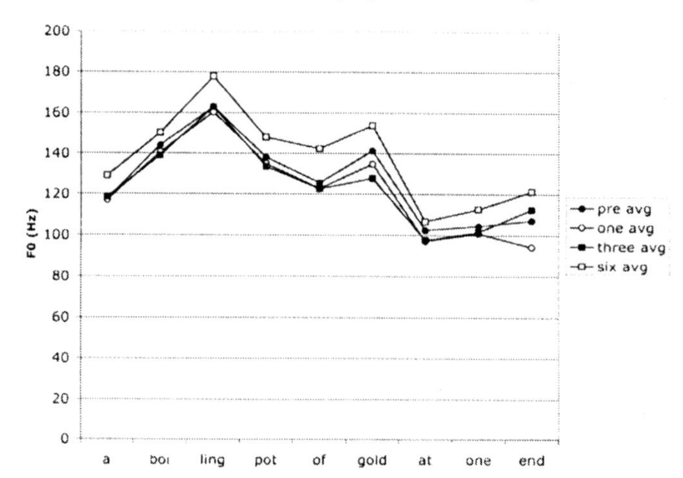

Figure C.8 Subject AM Rainbow Passage, phrase 8: Average F0

C.2 Subject DS

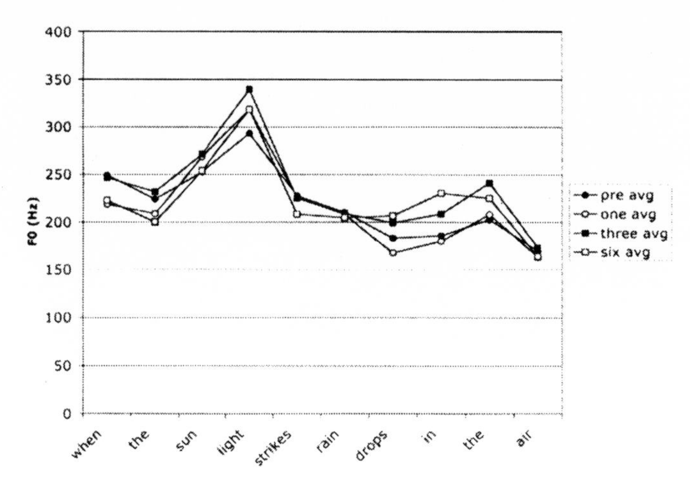

Figure C.12 Subject DS Rainbow Passage, phrase 1: Average F0

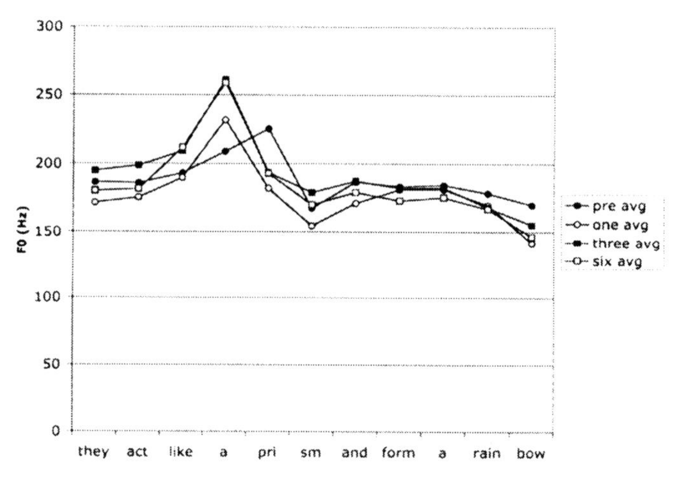

Figure C.13 Subject DS Rainbow Passage, phrase 2: Average F0

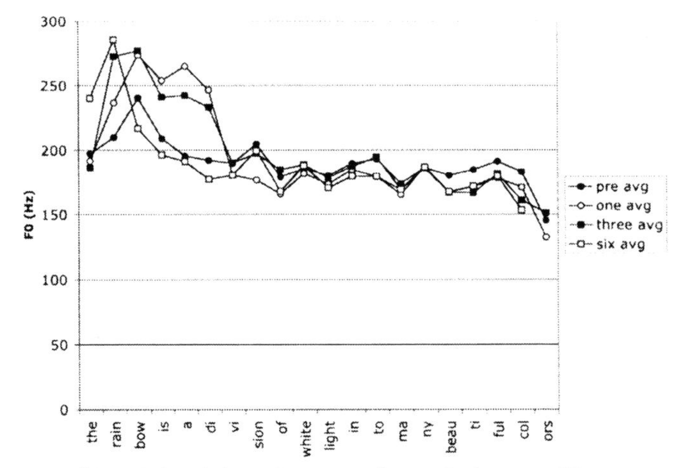

Figure C.14 Subject DS Rainbow Passage, phrase 3: Average F0

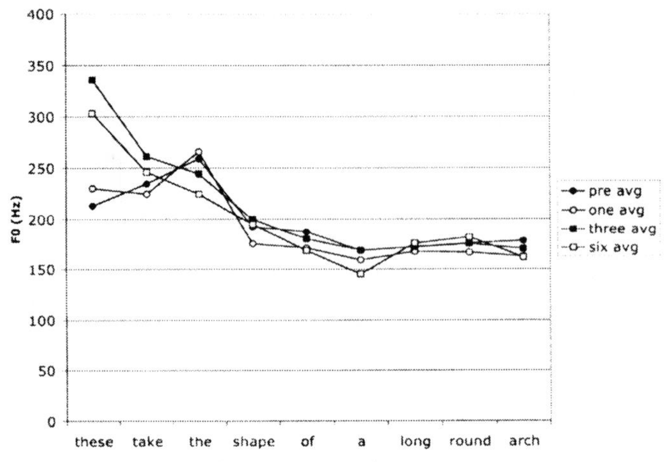

Figure C.15 Subject DS Rainbow Passage, phrase 4: Average F0

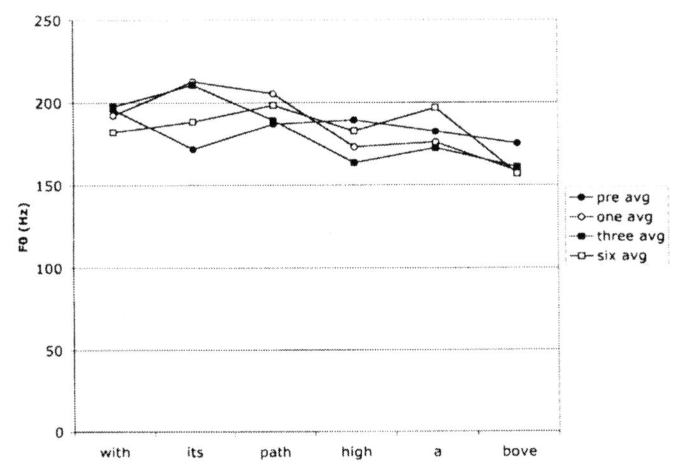

Figure C.16 Subject DS Rainbow Passage, phrase 5: Average F0

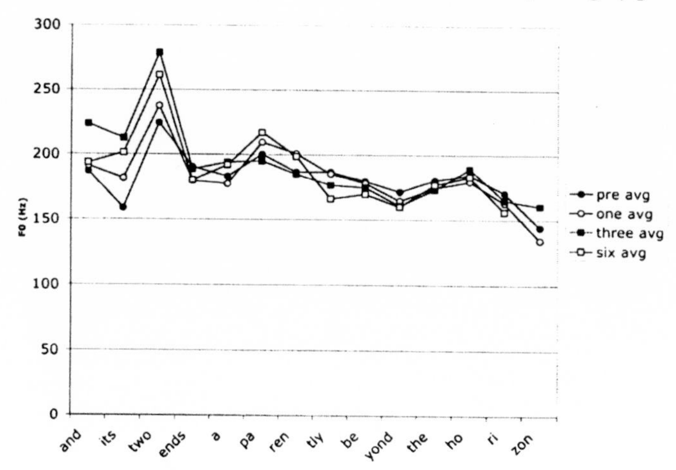

Figure C.17 Subject DS Rainbow Passage, phrase 6: Average F0

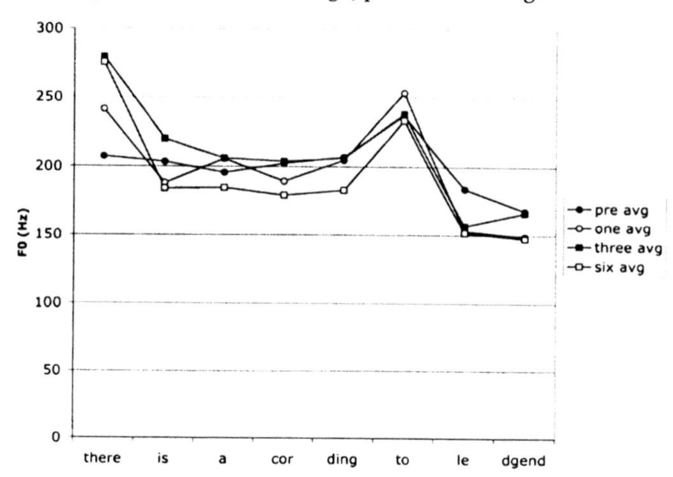

Figure C.18 Subject DS Rainbow Passage, phrase 7: Average F0

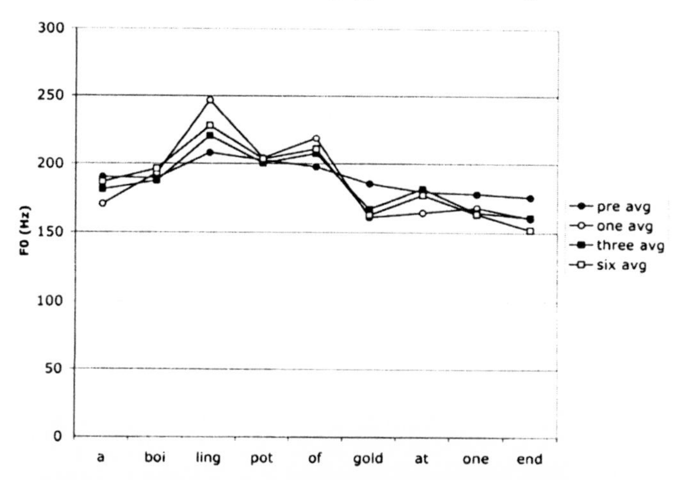

Figure C.19 Subject DS Rainbow Passage, phrase 8: Average F0

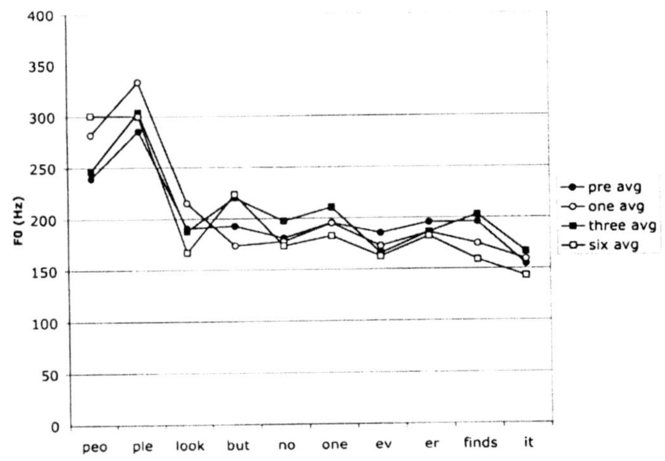

Figure C.20 Subject DS Rainbow Passage, phrase 9: Average F0

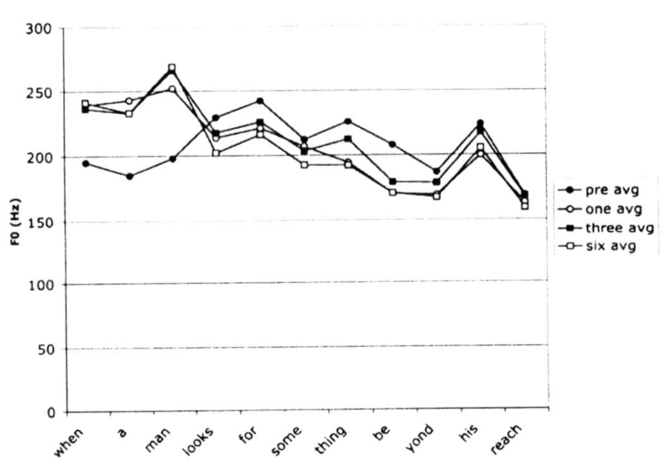

Figure C.21 Subject DS Rainbow Passage, phrase 10: Average F0

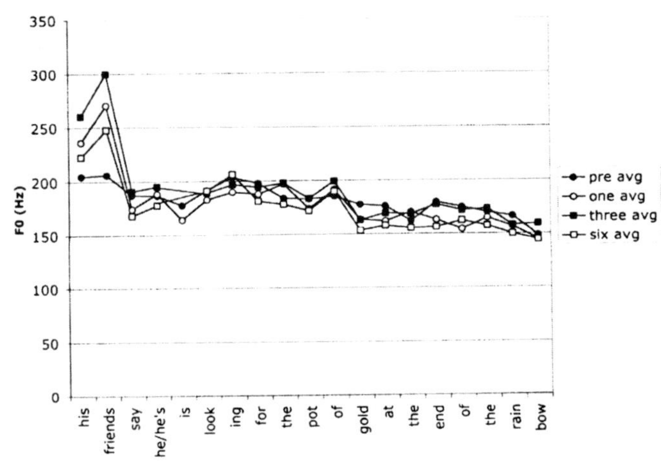

Figure C.22 Subject DS Rainbow Passage, phrase 11: Average F0

C.3 Subject MS

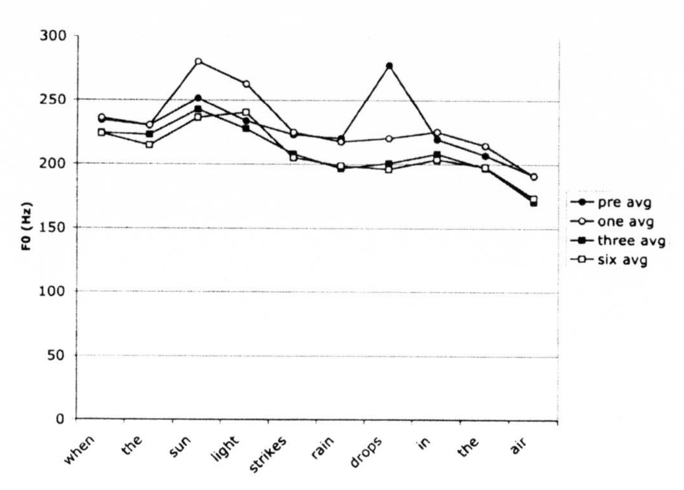

Figure C.23 Subject MS Rainbow Passage, phrase 1: Average F0

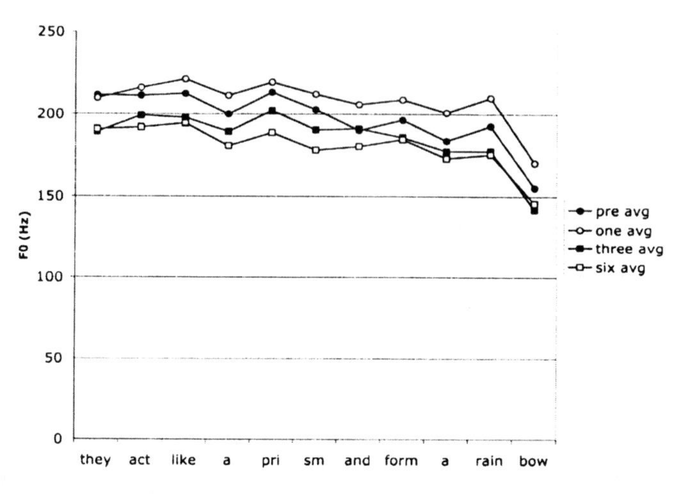

Figure C.24 Subject MS Rainbow Passage, phrase 2: Average F0

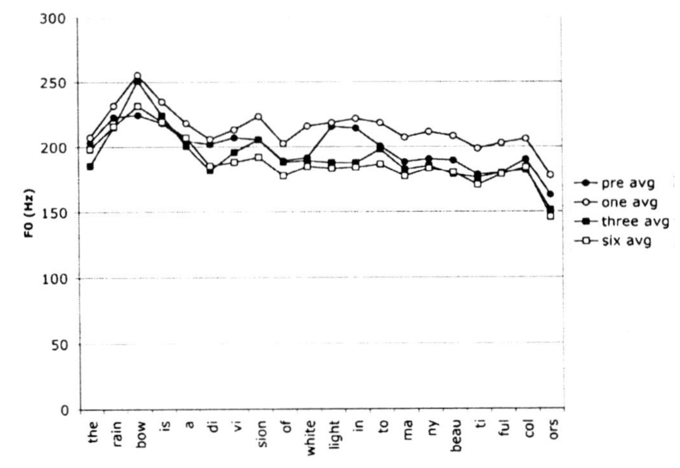

Figure C.25 Subject MS Rainbow Passage, phrase 3: Average F0

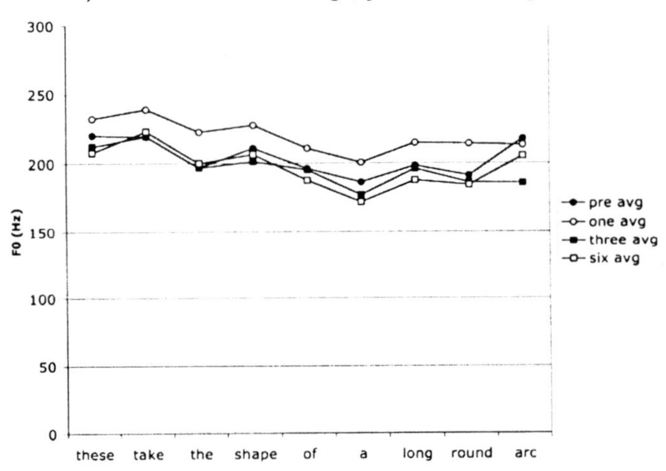

Figure C.26 Subject MS Rainbow Passage, phrase 4: Average F0

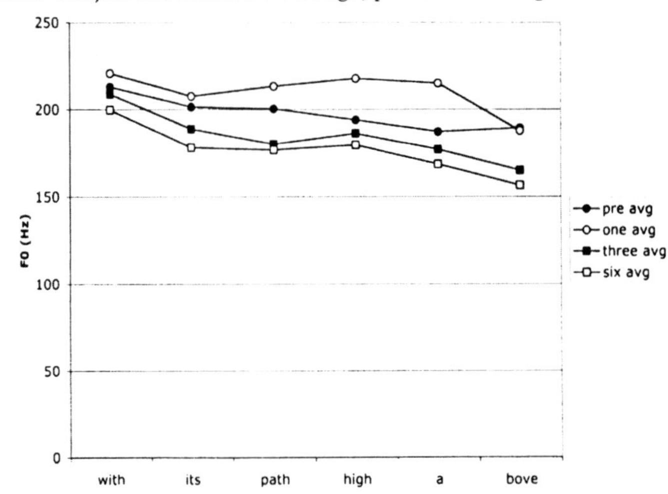

Figure C.27 Subject MS Rainbow Passage, phrase 5: Average F0

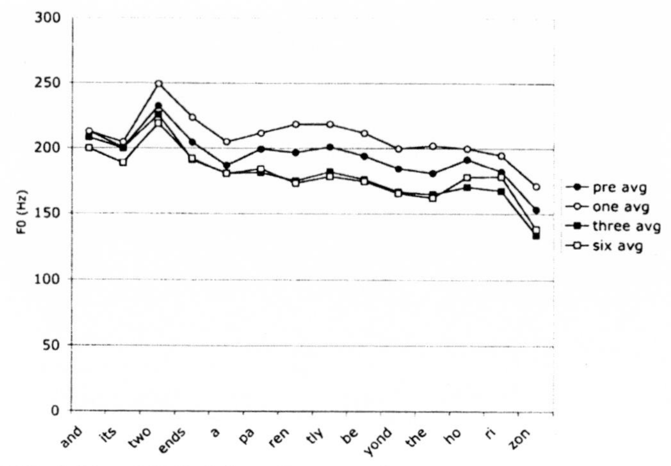

Figure C.28 Subject MS Rainbow Passage, phrase 6: Average F0

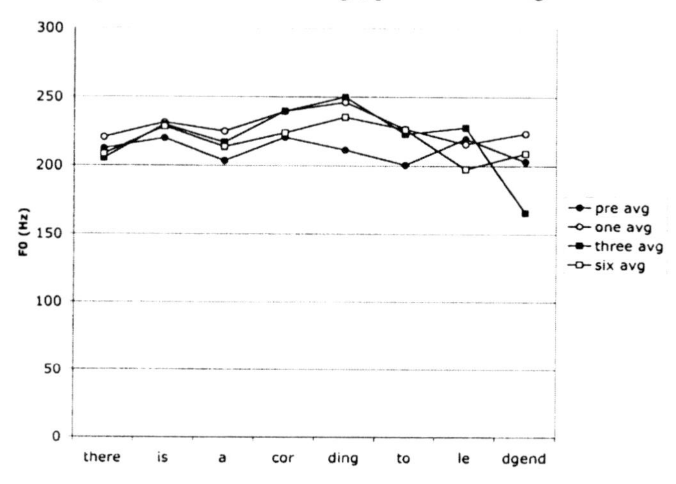

Figure C.29 Subject MS Rainbow Passage, phrase 7: Average F0

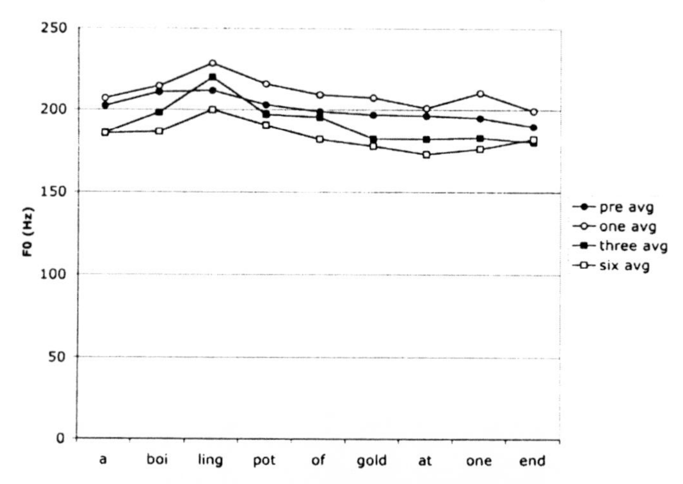

Figure C.30 Subject MS Rainbow Passage, phrase 8: Average F0

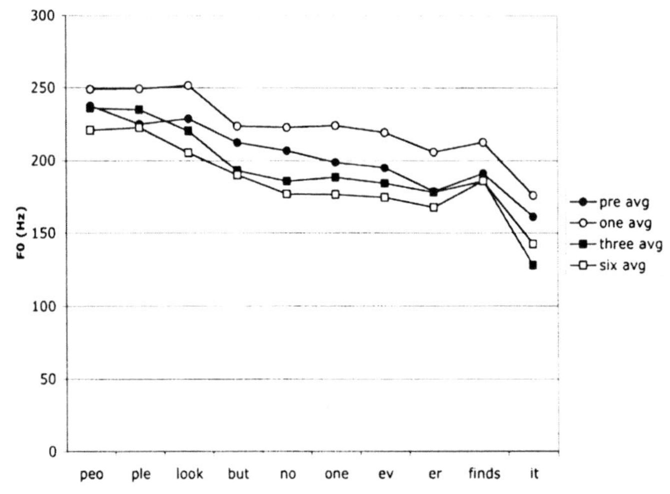

Figure C.31 Subject MS Rainbow Passage, phrase 9: Average F0

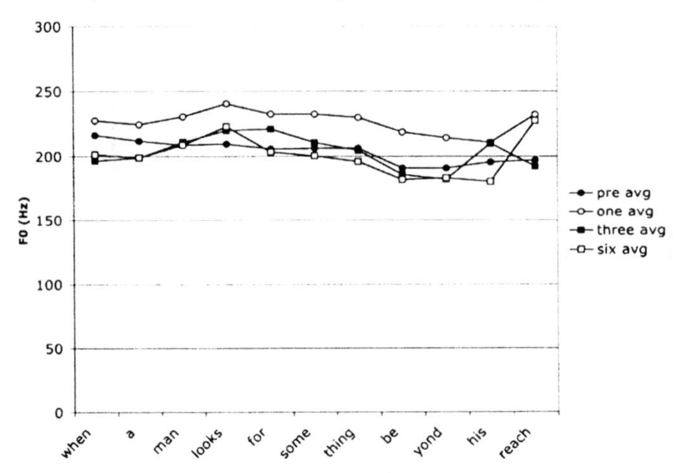

Figure C.32 Subject MS Rainbow Passage, phrase 10: Average F0

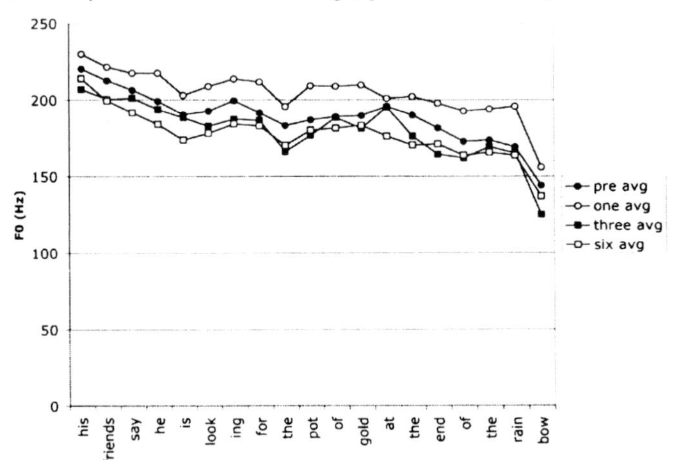

Figure C.33 Subject MS Rainbow Passage, phrase 11: Average F0

D Bev loves Bob figures

D.1 SUBJECT AM

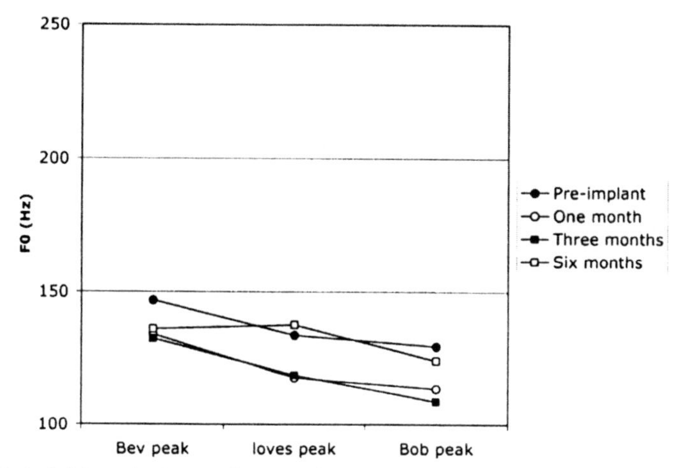

Figure D.1 Subject AM, Bev loves Bob. (average)

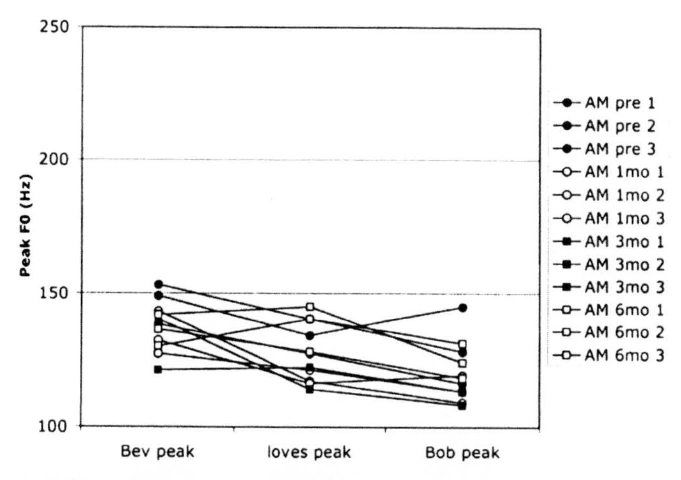

Figure D.2 Subject AM, Bev loves Bob.

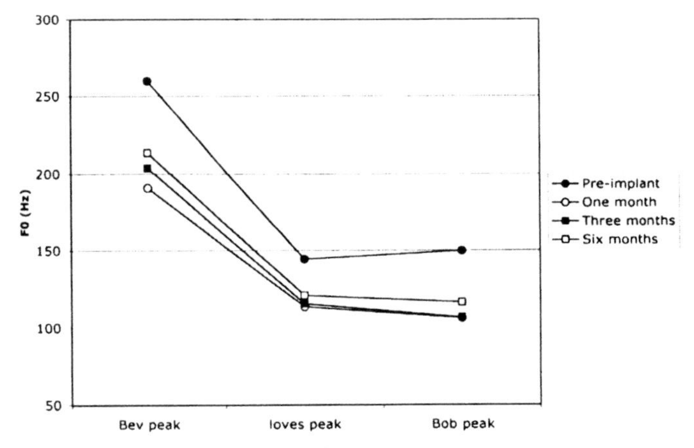

Figure D.3 Subject AM, Bev loves Bob. (average)

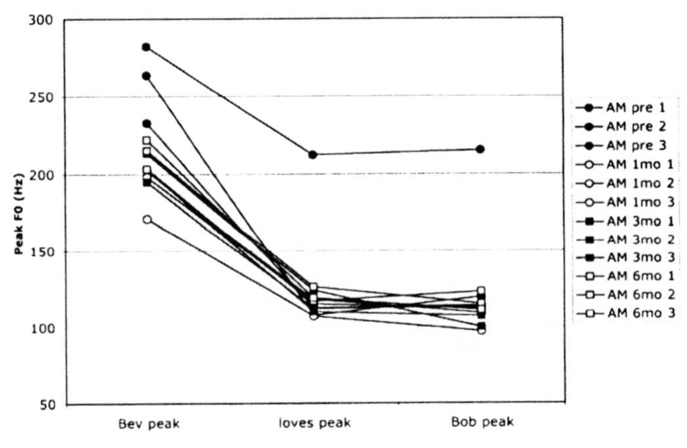

Figure D.4 Subject AM, Bev loves Bob.

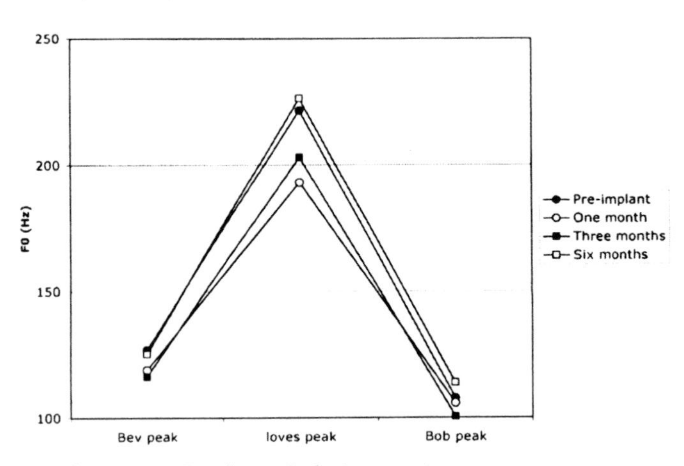

Figure D.5 Subject AM, Bev loves Bob. (average)

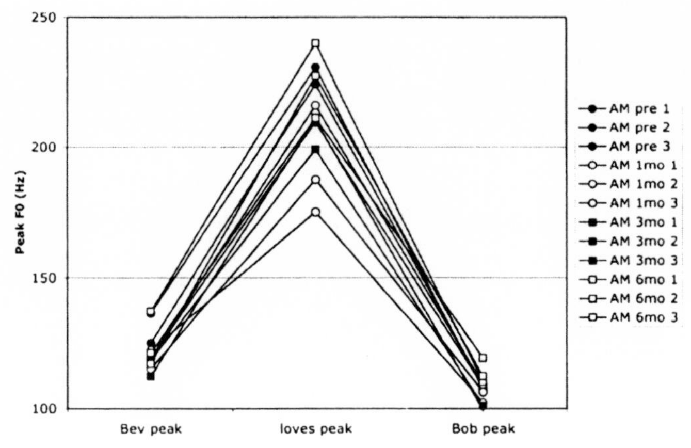

Figure D.6 Subject AM, Bev loves Bob.

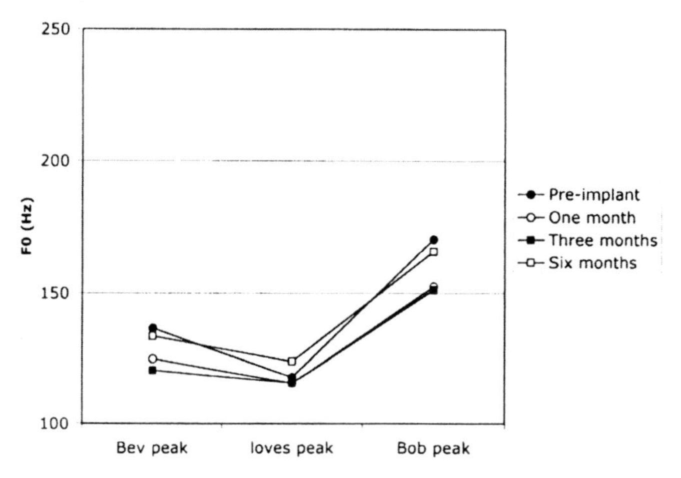

Figure D.7 Subject AM, Bev loves Bob. (average)

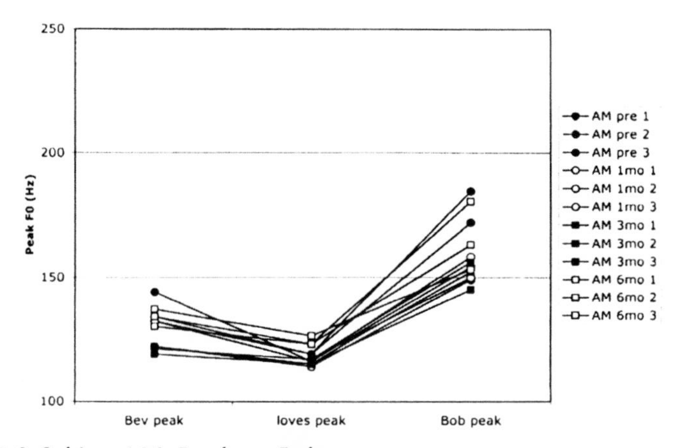

Figure D.8 Subject AM, Bev loves Bob.

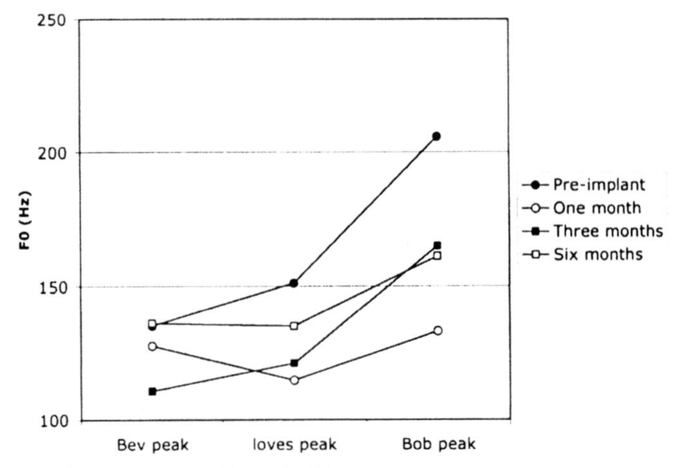

Figure D.9 Subject AM, Bev loves Bob? (average)

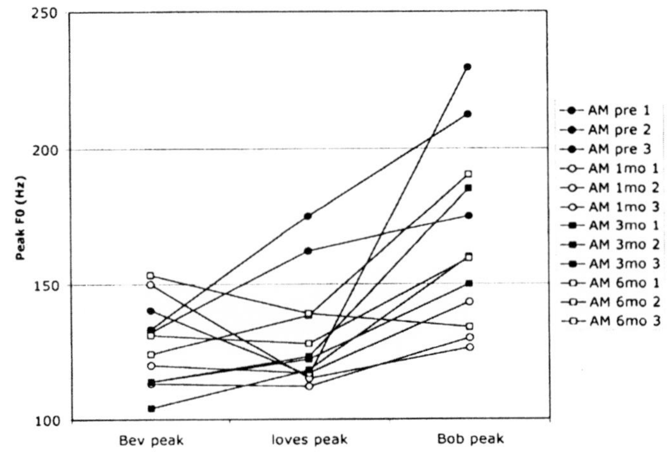

Figure D.10 Subject AM, Bev loves Bob?

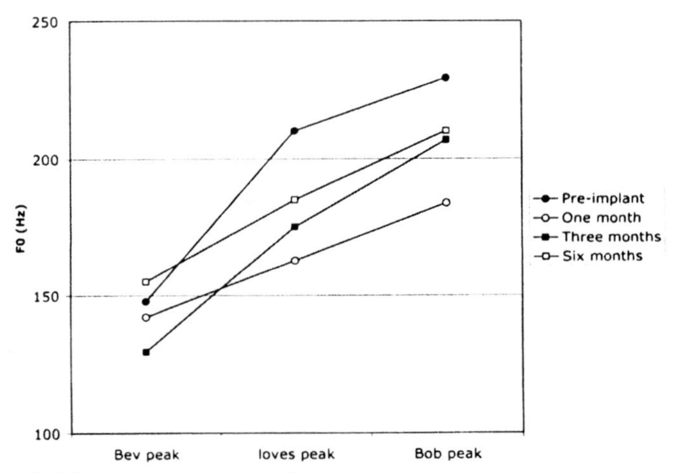

Figure D.11 Subject AM, <u>Bev</u> loves Bob? (average)

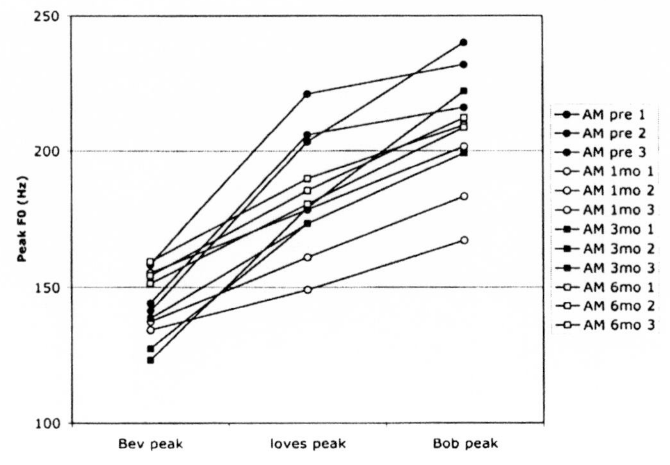

Figure D.12 Subject AM, <u>Bev</u> loves Bob?

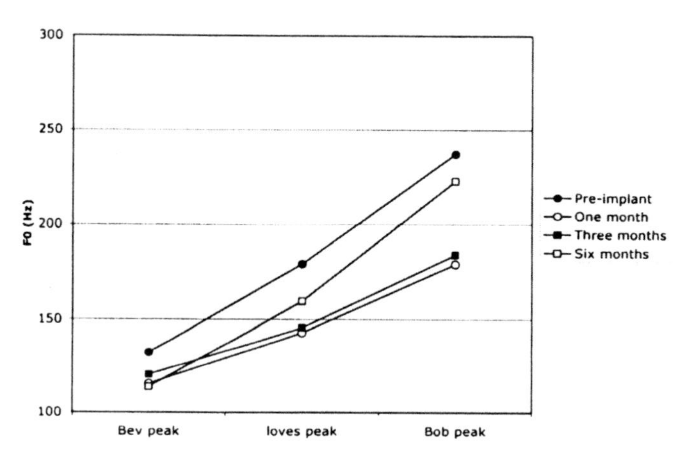

Figure D.13 Subject AM, Bev <u>loves</u> Bob? (average)

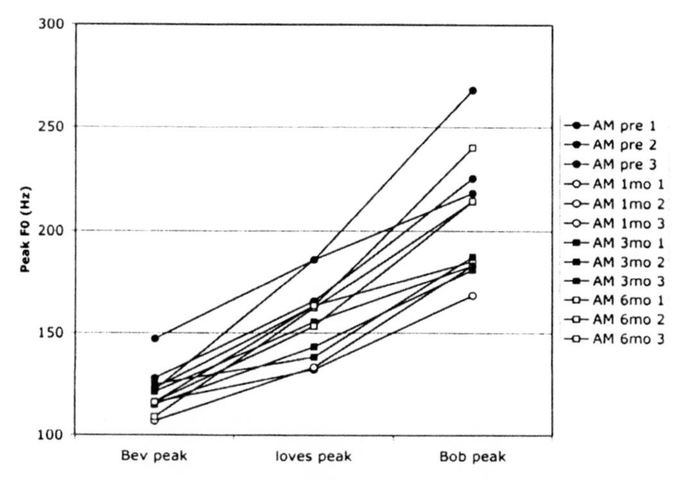

Figure D.14 Subject AM, Bev <u>loves</u> Bob?

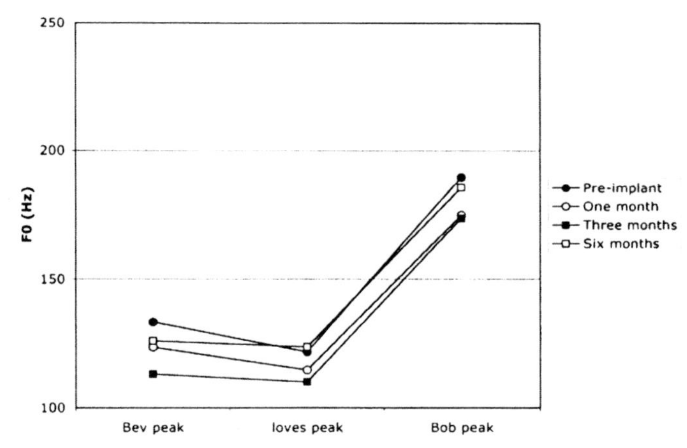

Figure D.15 Subject AM, Bev loves Bob? (average)

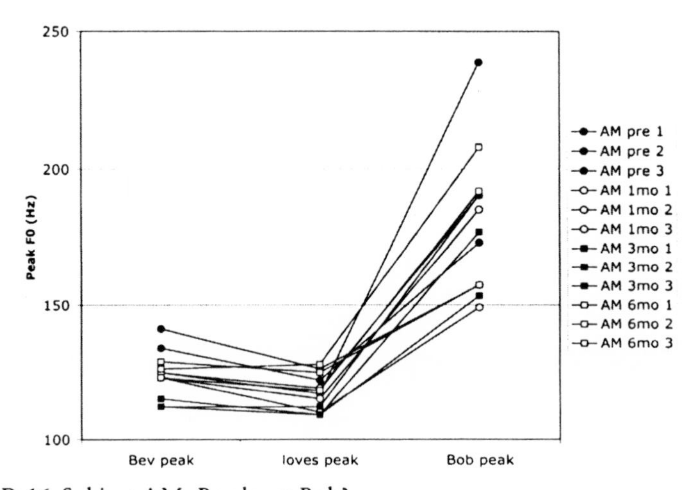

Figure D.16 Subject AM, Bev loves Bob?

D.2 SUBJECT DS

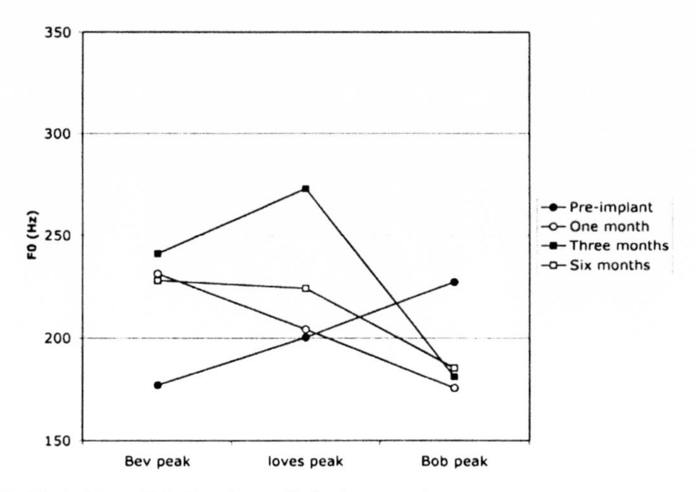

Figure D.17 Subject DS, Bev loves Bob. (average)

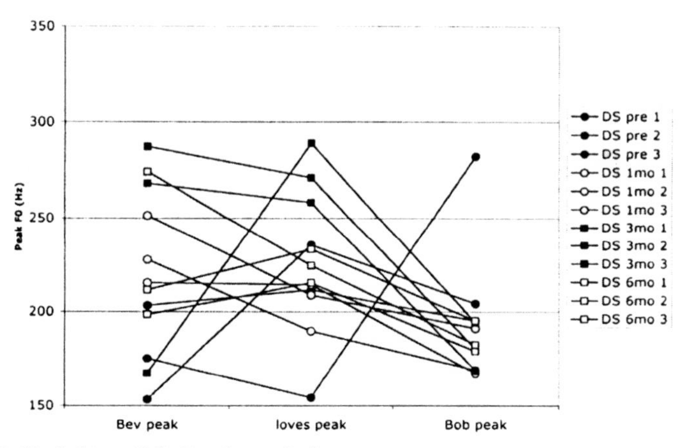

Figure D.18 Subject DS, Bev loves Bob.

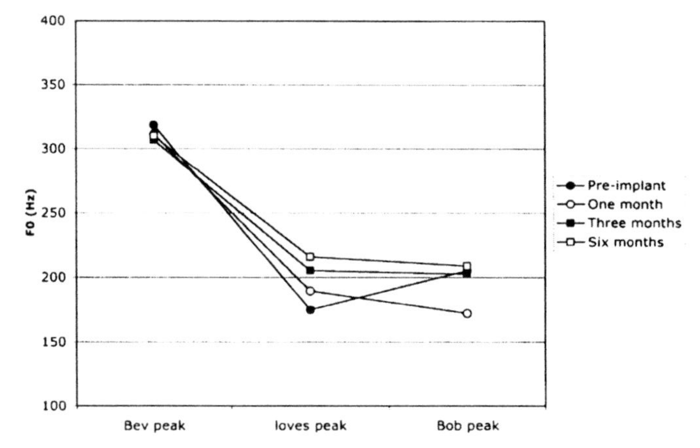

Figure D.19 Subject DS, Bev loves Bob. (average)

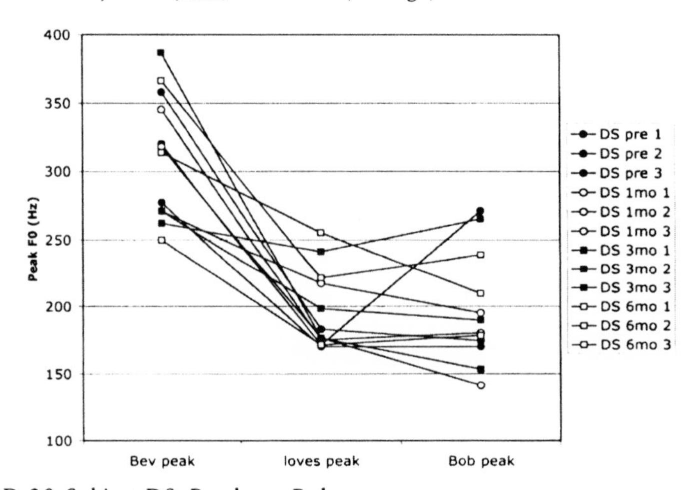

Figure D.20 Subject DS, Bev loves Bob.

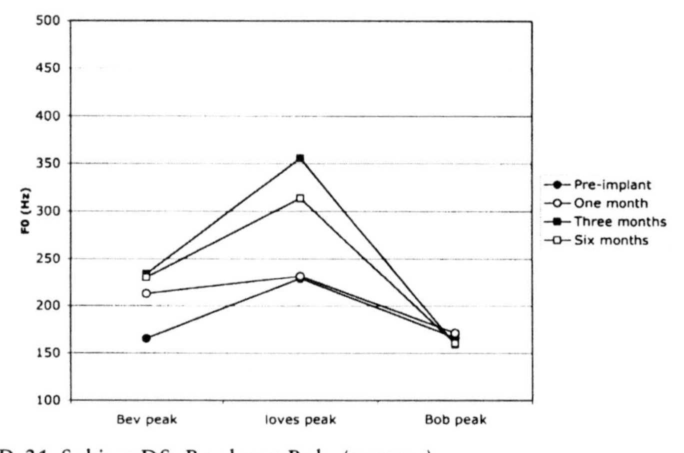

Figure D.21 Subject DS, Bev loves Bob. (average)

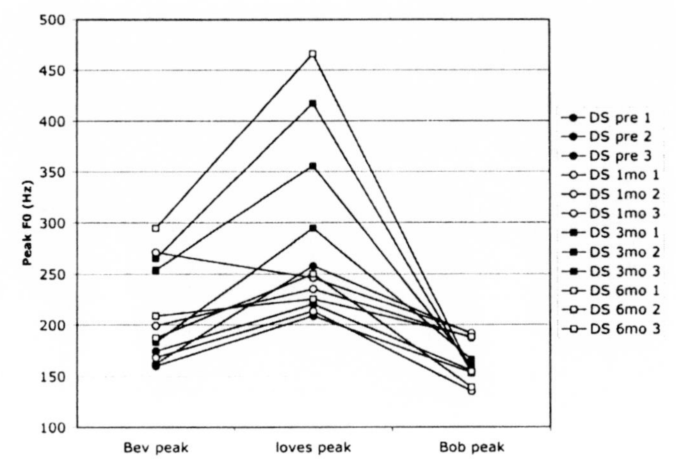

Figure D.22 Subject DS, Bev <u>loves</u> Bob.

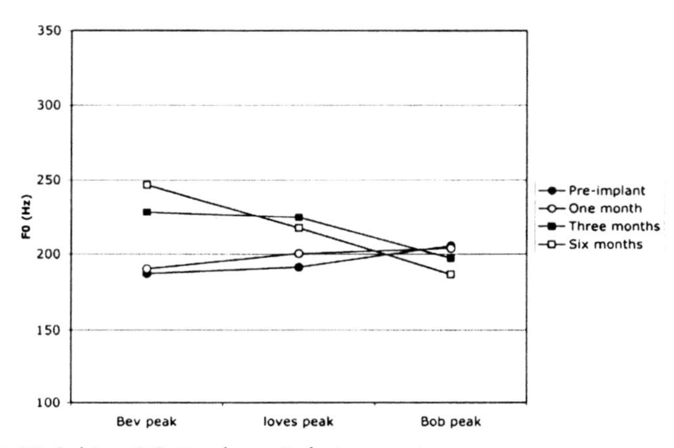

Figure D.23 Subject DS, Bev loves <u>Bob</u>. (average)

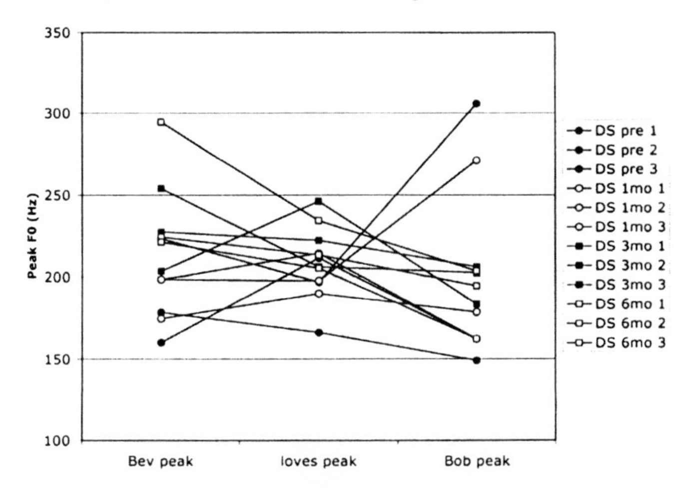

Figure D.24 Subject DS, Bev loves <u>Bob</u>.

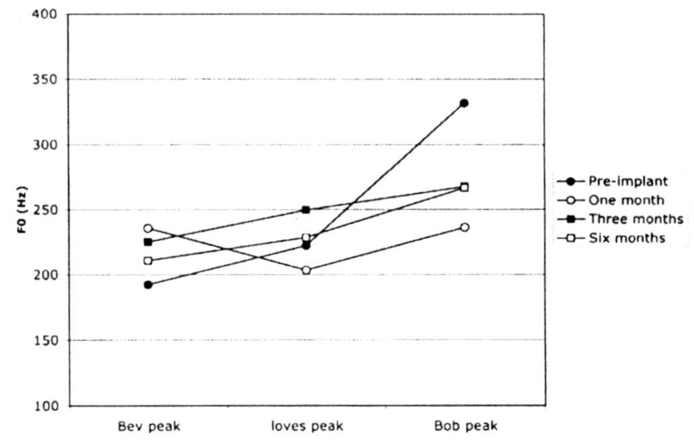

Figure D.25 Subject DS, Bev loves Bob? (average)

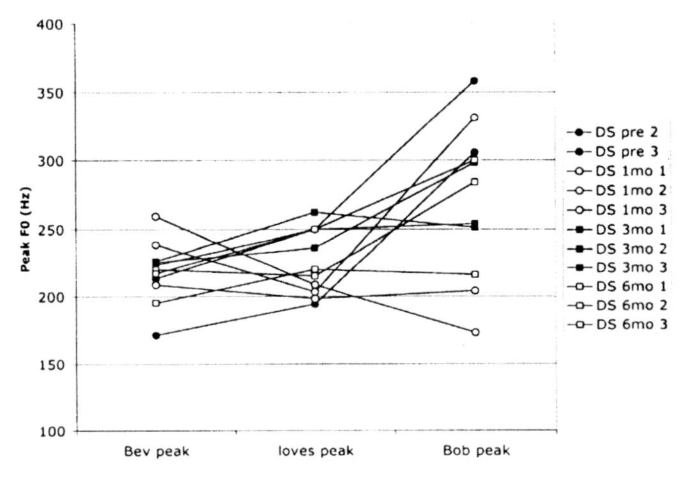

Figure D.26 Subject DS, Bev loves Bob?

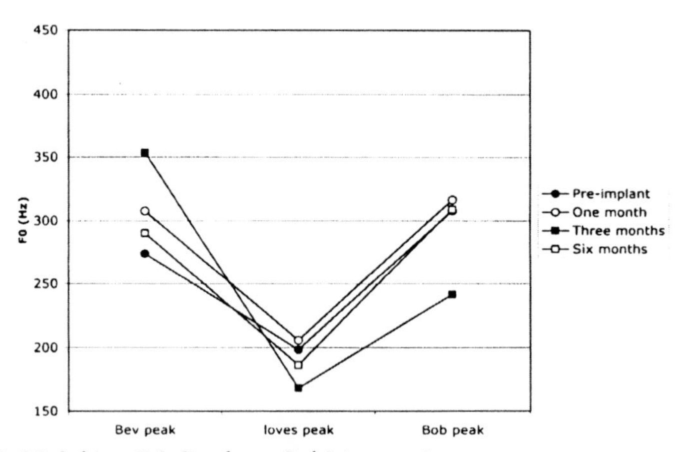

Figure D.27 Subject DS, <u>Bev</u> loves Bob? (average)

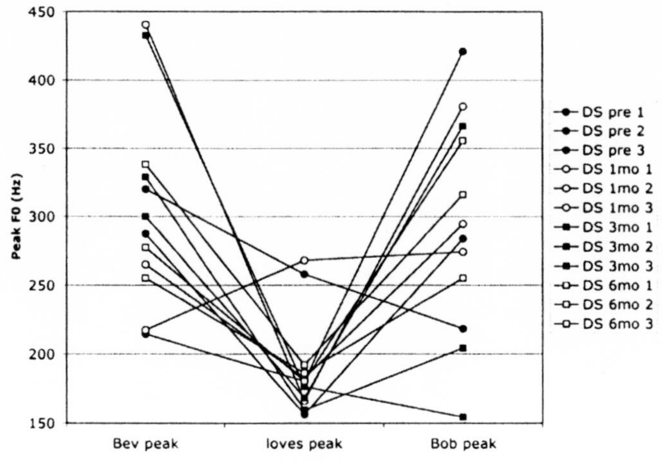

Figure D.28 Subject DS, <u>Bev</u> loves Bob?

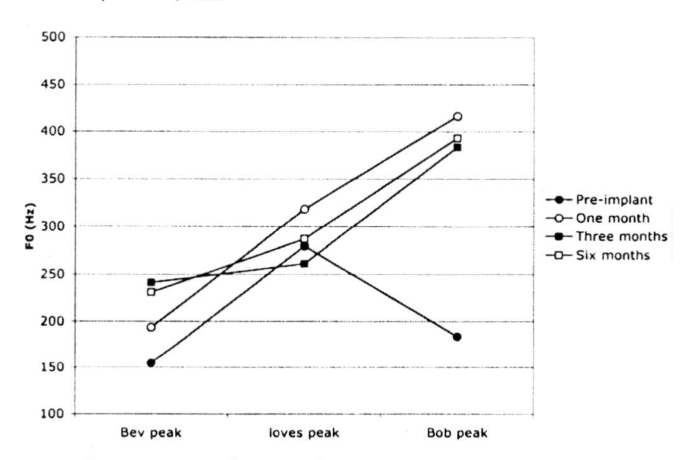

Figure D.29 Subject DS, Bev <u>loves</u> Bob? (average)

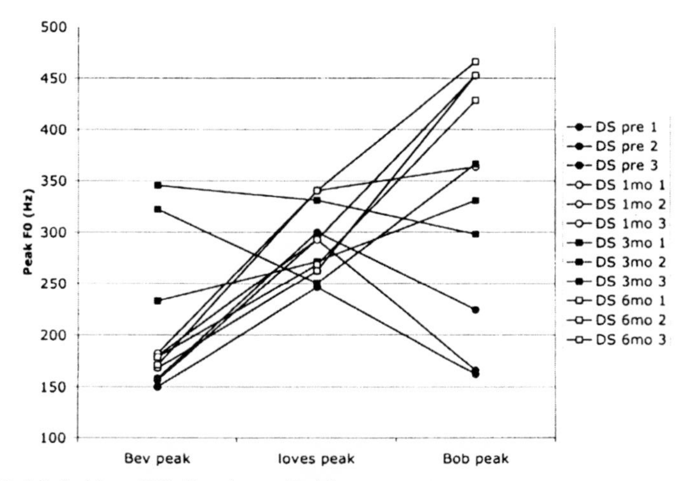

Figure D.30 Subject DS, Bev <u>loves</u> Bob?

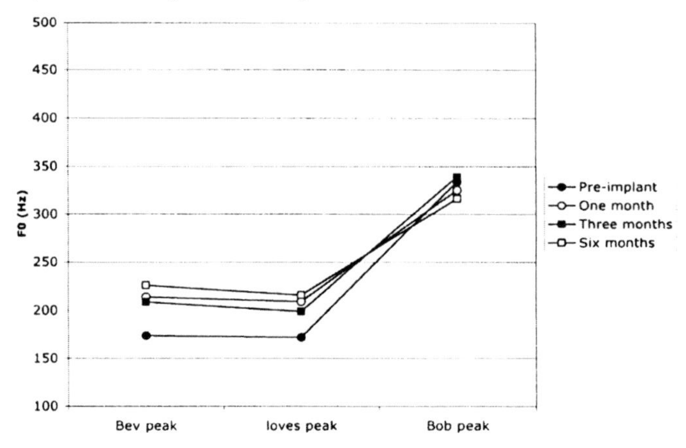

Figure D.31 Subject DS, Bev loves <u>Bob</u>? (average)

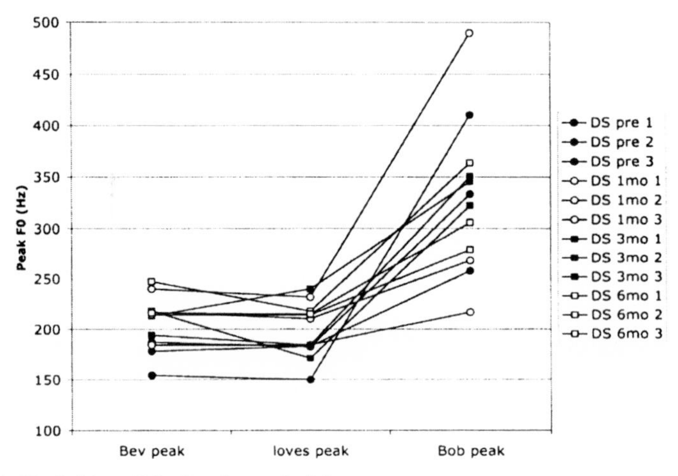

Figure D.32 Subject DS, Bev loves <u>Bob</u>?

D.3 SUBJECT MS

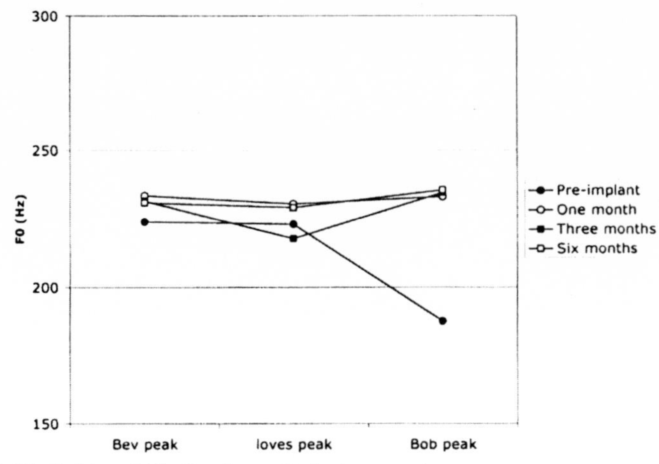

Figure D.33 Subject MS, Bev loves Bob. (average)

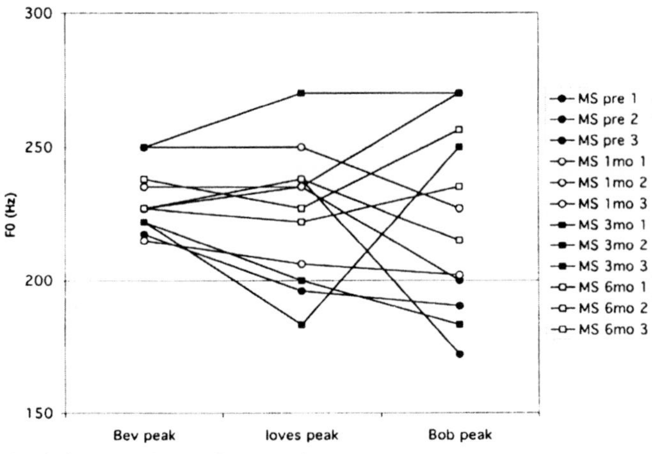

Figure D.34 Subject MS, Bev loves Bob.

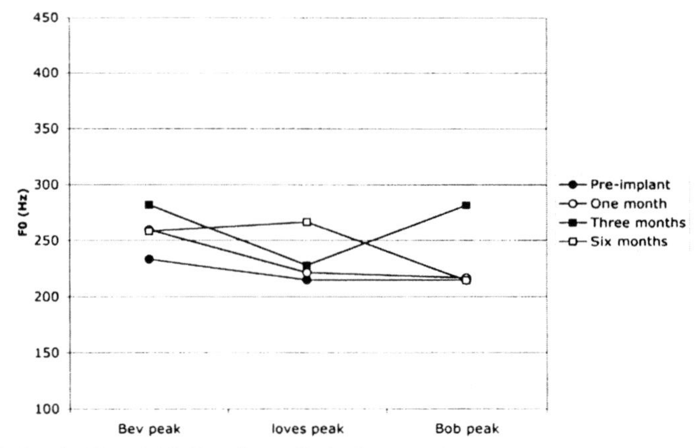

Figure D.35 Subject MS, Bev loves Bob. (average)

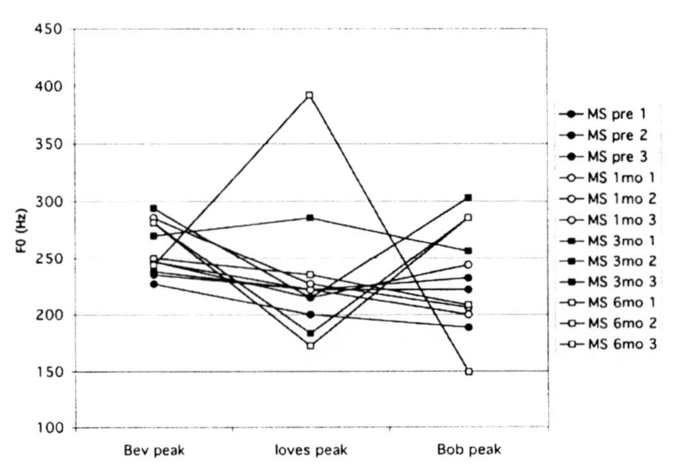

Figure D.36 Subject MS, Bev loves Bob.

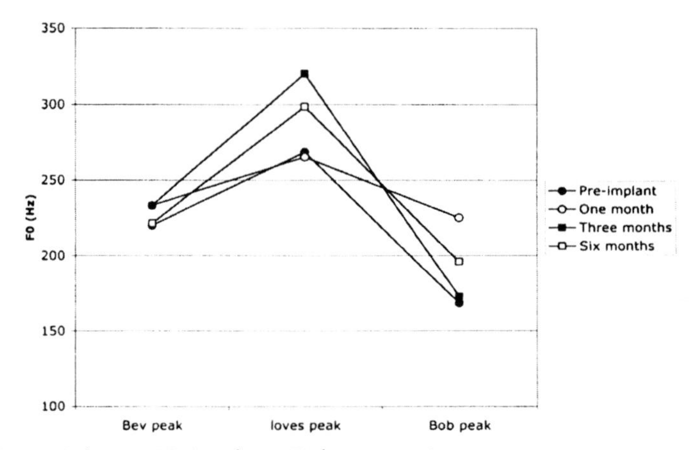

Figure D.37 Subject MS, Bev loves Bob. (average)

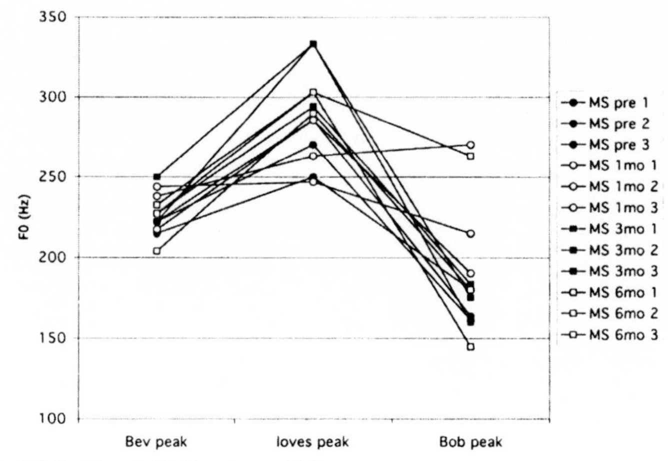

Figure D.38 Subject MS, Bev <u>loves</u> Bob.

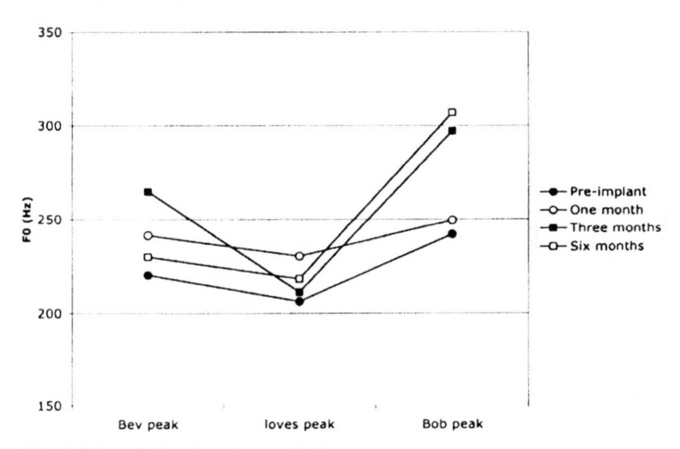

Figure D.39 Subject MS, Bev loves <u>Bob</u>. (average)

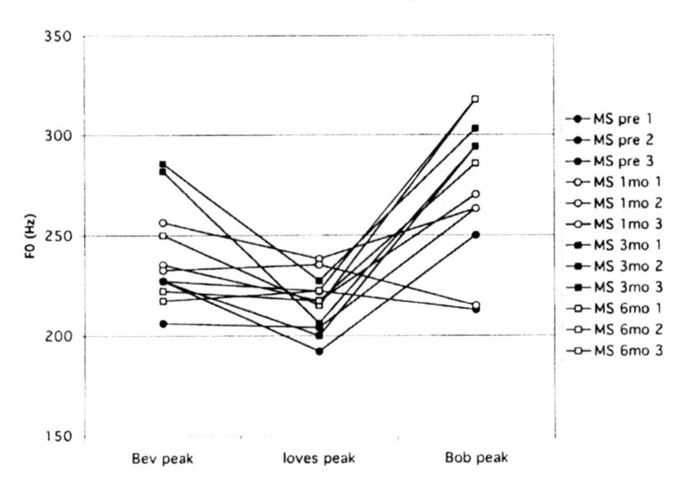

Figure D.40 Subject MS, Bev loves <u>Bob</u>.

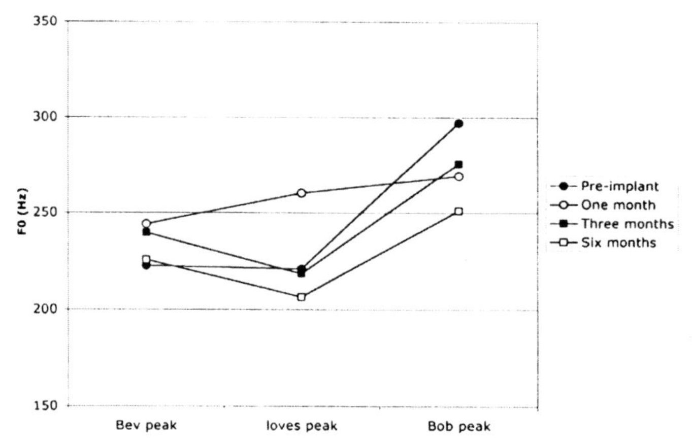

Figure D.41 Subject MS, Bev loves Bob? (average)

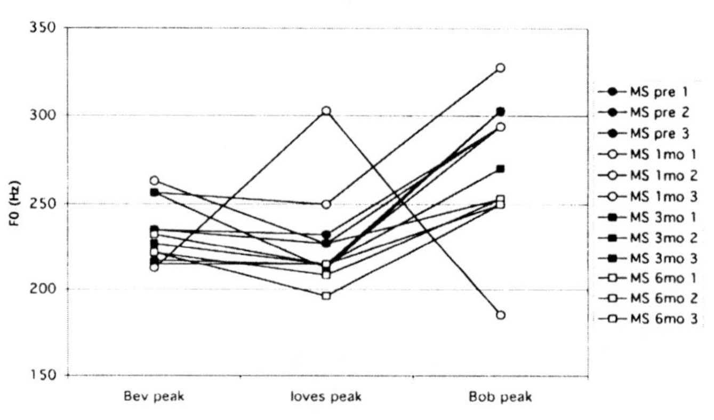

Figure D.42 Subject MS, Bev loves Bob?

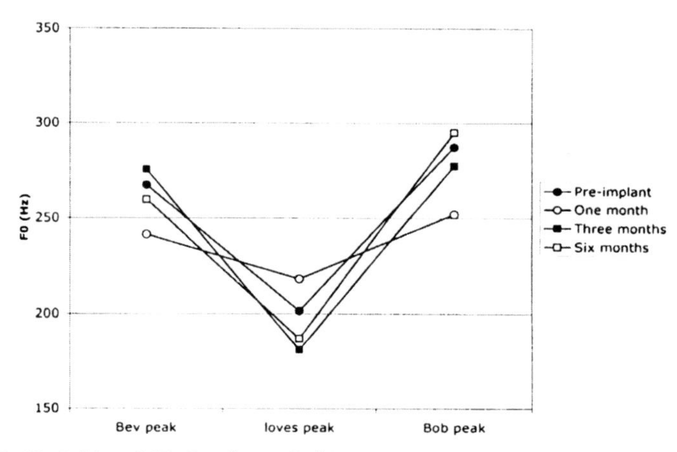

Figure D.43 Subject MS, <u>Bev</u> loves Bob? (average)

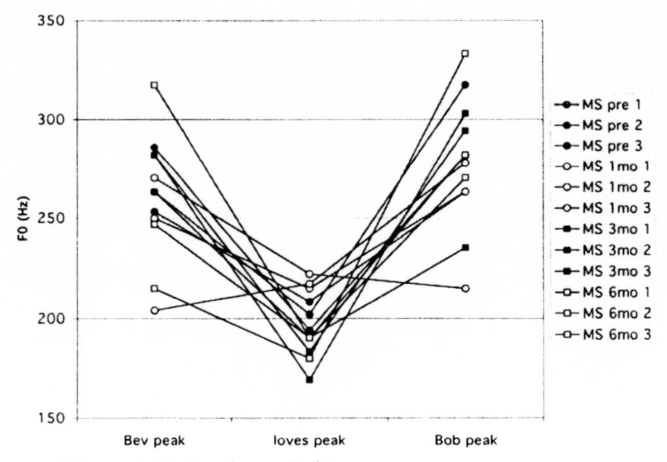

Figure D.44 Subject MS, <u>Bev</u> loves Bob?

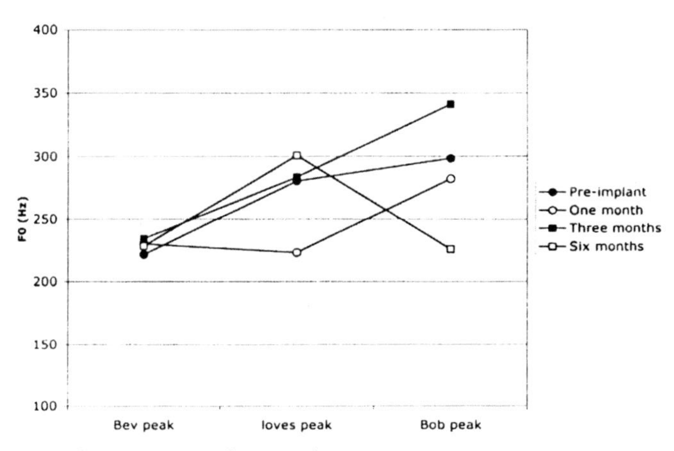

Figure D.45 Subject MS, Bev <u>loves</u> Bob? (average)

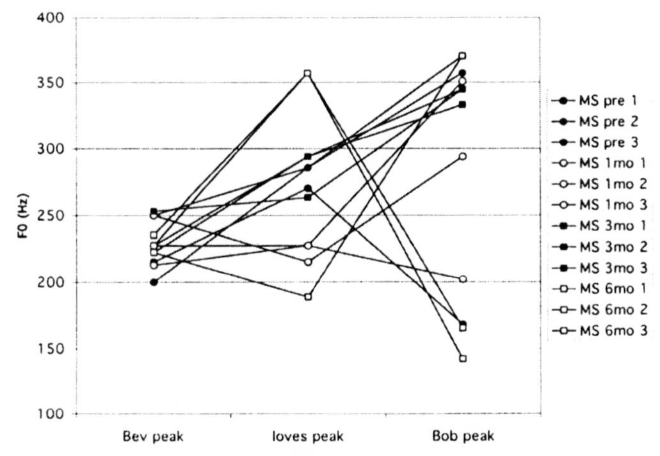

Figure D.46 Subject MS, Bev <u>loves</u> Bob?

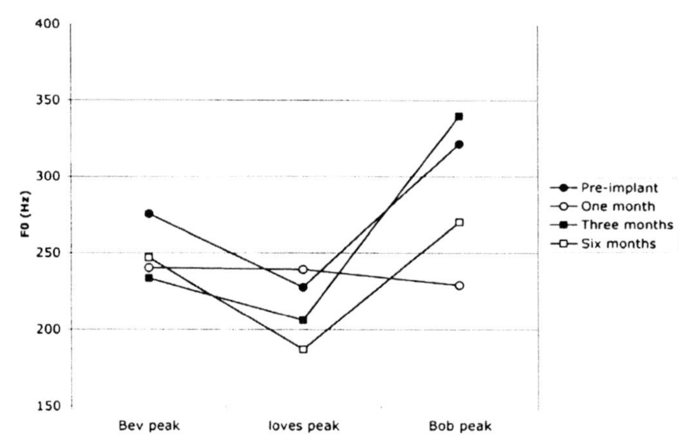

Figure D.47 Subject MS, Bev loves Bob? (average)

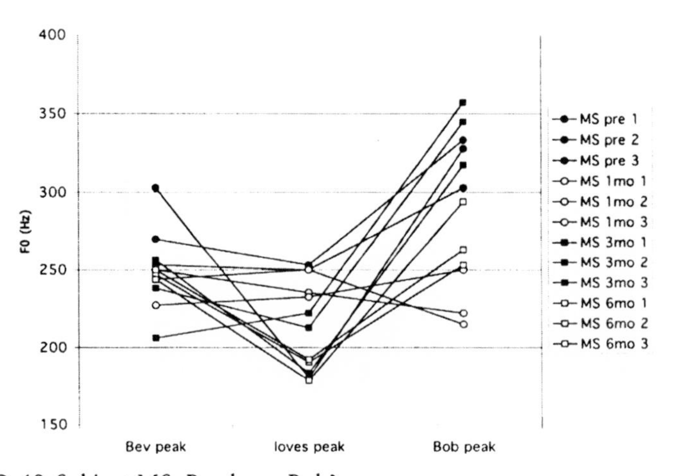

Figure D.48 Subject MS, Bev loves Bob?

Bibliography

Asker, Claes. 2001. Respiratory inductive plethysmograph (RIP). 9 May 2001. Linköpings Universitet. 4 June 2003. <http://www.imt.liu.se/bit/staff/ tomst/ rip.html>

Aronson, Josh. 2000. Panel discussion, Chicago International Film Festival. October 7, 2000.

Atkinson, J.R. 1973. Aspects of intonation in speech: Implications from an experimental study of fundamental frequency. Unpublished Ph.D. dissertation, University of Connecticut, Storrs. Cited in Lieberman and Blumstein (1988).

Biderman, Beverly. 1998. *Wired for Sound: A Journey Into Hearing.* Toronto: Trifolium Books Inc.

Blamey, Peter, Johanna Barry, Cathy Bow, Julia Sarant, Louise Paatsch, and Roger Wales. 2001. The development of speech production following cochlear implantation. *Clinical Linguistics & Phonetics* 15(5): 363–382.

Blamey, Peter J., Johanna G. Barry, and Pascale Jacq. 2001. Phonetic inventory development in young cochlear implant users 6 years postoperation. *Journal of Speech, Language, and Hearing Research* 44: 73–79.

Blamey, Peter J., Julia Z. Sarant, Louise E. Paatsch, Johanna G. Barry, Catherine P. Bow, Roger J. Wales, Maree Wright, Colleen Psarros, Kylie Rattigan, and Rebecca Tooher. 2001. Relationships Among Speech Perception, Production, Language, Hearing Loss, and Age in Children with Impaired Hearing. *Journal of Speech, Language, and Hearing Research* 44: 264–282.

Bradlow, A., G. Torretta, and D. Pisoni, "Intelligibility of normal speech I: Global and fine-grained acoustic-phonetic talker characteristics," *Speech Communication.* 20: 255–272, 1996.

Burnett, Theresa, Marcia B. Freedland, Charles R. Larson, and Timothy C. Hain. 1998. Voice F0 responses to manipulations in pitch feedback. *Journal of the Acoustical Society of America* 103(6): 3153–3161.

Carpenter, Mary. 1997. Learning to Hear Takes More Than Hardware; After Getting Cochlear Implants, Deaf Children in Special Programs Slowly Adapt to Sound. *Washington Post.* September 2, 1997. p. Z08.

Cheng, André K., Haya R. Rubin, Neil R. Powe, Nancy K. Mellon, Howard W. Francis, and John K. Niparko. 2000. Cost-Utility Analysis of the Cochlear Implant in Children. *Journal of the American Medical Association* 2000; 284: 850–856.

Chin, Steven B. 2002. Aspects of stop consonant production by pediatric users of cochlear implants. *Language, Speech, and Hearing Services in Schools* 33: 38–51.

Chin, Steven B., Kevin R. Finnegan, and Brian A. Chung. 2001. Relationships among types of speech intelligibility in pediatric users of cochlear implants. *Journal of Communication Disorders* 34: 187–203.

CII Bionic Ear Implant. 2003. Advanced Bionics Corporation. <http://www.cochlearimplant.com/products/cii_be_implant.html> 9 June 2003.
Ciocca, Valter, Alexander L. Francis, Rani Aisha, and Lena Wong. 2002. The perception of Cantonese lexical tones by early-deafened cochlear implantees. *Journal of the Acoustical Society of America* 111(5): 2250–2256.
Cochlear Corporation. 1999. *Issues and Answers: The Nucleus® 24 Cochlear Implant System*.
Cowie, R., E. Douglas-Cowie, and A.G. Kerr. 1984. A study of speech deterioration in post-lingually deafened adults. *Journal of Laryngology and Otology* 96: 101–112.
Dorman, Michael F., Philipos C. Loizou, Anthony J. Spahr, and Erin Maloff. 2002. A Comparison of the Speech Understanding Provided by Acoustic Models of Fixed-Channel and Channel-Picking Signal Processors for Cochlear Implants. *Journal of Speech, Language, and Hearing Research* 45: 783–788.
Ecklund, Amy. 1999. Now Hear This! Deaf since childhood, a TV actress celebrates the gift of hearing. *People Weekly*. July 12, 1999. p. 67–70.
Economou, Alexandra, Vivien C. Tartter, Patricia M. Chute and Sharon A. Hellman. 1992. Speech changes following reimplantation from a single-channel to a multichannel cochlear implant. *Journal of the Acoustical Society of America* 92(3): 1310–1323.
Elman, Jeffrey L. 1981. Effects of frequency-shifted feedback on the pitch of vocal productions. *Journal of the Acoustical Society of America* 70(1): 45–50.
Erenberg, Sheryl. 2001. Personal communication, January 29, 2001.
Ertmer, David J. 2001. Emergence of a vowel system in a young cochlear implant recipient. *Journal of Speech, Language, and Hearing Research* 44: 803–813.
Fairbanks, G. 1960. *Voice and articulation drillbook*. New York: Harper.
Fourakis, M., M. Skinner, T. Holden, L. Holden. 2001. Evaluation of word recognition performance by recipients of the Nucleus 24 cochlear implant. Abstract from the *New Frontiers in the Amelioration of Hearing Loss* meeting program, p. 32.
Friesen, Lendra M, Robert V. Shannon, Deniz Baskent, and Xiaosong Wang. 2001. Speech recognition in noise as a function of the number of spectral channels: Comparison of acoustic hearing and cochlear implants. *Journal of the Acoustical Society of America* 110(2): 1150–1163.
Fryauf-Bertschy, Holly, Richard S. Tyler, Danielle M. Kelsay, and Bruce J. Gantz. 1992. Performance over Time of Congenitally Deaf and Postlingually Deafened Children Using a Multichannel Cochlear Implant. *Journal of Speech and Hearing Research* 35(4): 913–20.
Fryauf-Bertschy, Holly, Richard S. Tyler, Danielle M. Kelsay, Bruce J. Gantz, and George G. Woodworth. 1997. Cochlear Implant Use by Prelingually Deafened Children: The Influences of Age at Implant and Length of Device Use. *Journal of Speech, Language, and Hearing Research* 40: 183–199.
Fu, Qian-Jie and John J. Galvin III. 2001. Recognition of spectrally asynchronous speech by normal-hearing listeners and Nucleus-22 cochlear implant users. *Journal of the Acoustical Society of America* 109(3): 1166–1172.
Fu, Qian-Jie, John J. Galvin III, and Xiaosong Wang. 2001. Recognition of time-distorted sentences by normal-hearing and cochlear-implant listeners. *Journal of the Acoustical Society of America* 109(1): 379–384.
Garber, Sharon B. and Karlind T. Moller. 1979. The effects of feedback filtering on nasalization in normal and hypernasal speakers. *Journal of Speech and Hearing Research* 22: 321–333.
Geers, Ann E. 2002. Factors affecting the development of speech, language, and literacy in children with early cochlear implantation. *Language, Speech, and Hearing Services in Schools* 33: 172–183.

Geers, Ann, Brent Spehar, and Allison Sedey. 2002. Use of speech by children from total communication programs who wear cochlear implants. *American Journal of Speech-Language Pathology* 11: 50–58.

Geurts, Luc and Jan Wouters. 2001. Coding of the fundamental frequency in continuous interleaved sampling processors for cochlear implants. *Journal of the Acoustical Society of America* 109(2): 713–726.

Goehl, Henry and Diana K. Kaufman. 1984. Do the effects of adventitious deafness include disordered speech? *Journal of Speech and Hearing Disorders.* 49: 58–64.

Goffman, Lisa, David J. Ertmer and Christa Erdle. 2002. Changes in speech production in a child with a cochlear implant: Acoustic and kinematic evidence. *Journal of Speech, Language, and Hearing Research* 45: 891–901.

Gurchiek, Kathy. 1998. Hearing-impaired getting help from cochlear implants. *The Associated Press State and Local Wire*, December 19, 1998, Saturday, BC cycle.

Hamzavi, Jafar, Wolf Dieter Baumgartner, and Stefan Marcel Pok. 2001. Does cochlear reimplantation affect speech recognition? *International Journal of Audiology.* 41: 151–156.

Helms, Müller, Schön, Winkler, Moser, Shehata-Dieler, Kastenbauer, Baumann, Rasp, Schorn, Esser, Baumgartner, Hamzavi, Gstöttner, Westhofen, Döring, Dujardin, Albegger, Mair, Zenner, Haferkamp, Schmitz-Salue, Arold, Sesterhenn, Jahnke, Wagner, Gräbel, Bockmühl, Häusler, Vischer, Kompis, Hildmann, Radü, Stark, Engel, Hildmann, Streitberger, Hüttenbrink, Müller-Aschoff, Hoffman, Seeling, Hloucal, von Ilberg, Kiefer, Pfennigdorf, Gall, Breitfuss, Stelzig, Begall, Hey, Vorwerk, Thumfart, Gunkel, Zorowka, Stephan, Gammert, Mathis, DeMin, Freigang, Ziese, Stützel, von Specht, Arnold, Brockmeier, Ebenhoch, Steinhoff, Zierhofer, Zwicknagl, and Stöbig. 2001. Comparison of the TEMPO+ Ear-Level Speech Processor and the CIS PRO+ Body-Worn Processor in Adult MED-EL Cochlear Implant Users. *Annals of Otology, Rhinology, and Laryngology.* 63: 31–40.

Hopkins 24/7. 2001. ABC News. 19 April 2001 <http://abcnews.go.com/onair/hopkins/>.

Ingram, David. 2002. The measurement of whole-word productions. *Journal of Child Language* 29: 713–733.

Kessler, Dorcas K. 1999. The Clarion® Multi-Strategy Cochlear Implant. *Annals of Otology, Rhinology, and Laryngology.* 108(4) part 2: 8–16.

Kirk, Karen Iler, Nancy Tye-Murray, and Richard R. Hurtig. 1992. The use of static and dynamic vowel cues by multichannel cochlear implant users. *Journal of the Acoustical Society of America* 91(6): 3487–98.

Kishon-Rabin, Liat, Riki Taitelbaum, Yishai Tobin, and Minka Hildesheimer. 1999. The effect of partially restored hearing on speech production of postlingually deafened adults with multichannel cochlear implants. *Journal of the Acoustical Society of America* 106(5): 2843–2857.

Lane, Harlan, Robert Hoffmeister, and Ben Bahan. 1996. *A Journey into the Deaf-World.* San Diego: DawnSignPress.

Lane, Harlan, Melanie Matthies, Joseph Perkell, Jennell Vick, and Majid Zandipour. 2001. The effects of changes in hearing status in cochlear implant users on the acoustic vowel space and CV coarticulation. *Journal of Speech, Language, and Hearing Research* 44: 552–563.

Lane, Harlan, and Bernard Tranel. 1971. The Lombard sign and the role of hearing in speech. *Journal of Speech and Hearing Research* 14: 677–709.

Lane, Harlan and Jane Wozniak Webster. 1991. Speech deterioration in postlingually deafened adults. *Journal of the Acoustical Society of America* 89(2): 859–866.

Lane, Harlan, Jane Wozniak, Melanie Matthies, Mario Svirsky, and Joseph Perkell. 1995. Phonemic Resetting vs. Postural Adjustments in the Speech of Cochlear Implant Users: An Exploration of Voice-Onset Time. *Journal of the Acoustical Society of America* 98(6): 3096–3106.
Lane, Harlan, Jane Wozniak, Melanie Matthies, Mario Svirsky, Joseph Perkell, Michael O'Connell, and Joyce Manzella. 1997. Changes in sound pressure and fundamental frequency contours following changes in hearing status. *Journal of the Acoustical Society of America* 101(4): 2244–2252.
Lane, Harlan, Jane Wozniak, and Joseph Perkell. 1994. Changes in Voice-Onset Time in Speakers with Cochlear Implants. *Journal of the Acoustical Society of America* 96(1): 56–64.
Langereis, M.C., A.J. Bosman, A.F. van Olphen, and G.F. Smoorenburg. 1995. Changes in Vowel Quality in Adult Cochlear Implant Users. *Annals of Otology, Rhinology, and Laryngology.* 104(9) part 2: 387–390.
Langereis, Margreet C., Arjan J. Bosman, Adriaan F. van Olphen, and Guido F. Smoorenburg. 1997. Changes in Vowel Quality in Post-lingually Deafened Cochlear Implant Users. *Audiology* 36: 279–297.
Langereis, Margreet C., Arjan J. Bosman, Adriaan F. van Olphen, and Guido F. Smoorenburg. 1998. Effect of Cochlear Implantation on Voice Fundamental Frequency in Post-lingually Deafened Adults. *Audiology* 37: 219–230.
Langereis, M.C., P.H. Dejonckere, A.F. van Olphen, G.F. Smoorenberg. 1997. Effect of Cochlear Implantation on Nasality in Post-Lingually Deafened Adults. *Folia Phoniatrica Logopedia.* 49: 308–314.
Lieberman, Philip, and Sheila E. Blumstein. 1988. *Speech physiology, speech perception, and acoustic phonetics.* Cambridge University Press.
Loizou, Philipos C., Michael F. Dorman, and Verelle Powell. 1998. The recognition of vowels produced by men, women, boys, and girls by cochlear implant patients using a six-channel CIS processor. *Journal of the Acoustical Society of America* 103(2): 1141–1149.
Loizou, Philipos C and Oguz Poroy. 2001. Minimum spectral contrast needed for vowel identification by normal hearing and cochlear implant listeners. *Journal of the Acoustical Society of America* 110(3): 1610–1627.
Matthies, Melanie L, Mario A. Svirsky, Harlan L. Lane and Joseph S. Perkell. 1994. A Preliminary Study of the Effects of Cochlear Implants on the Production of Sibilants. *Journal of the Acoustical Society of America* 96(3): 1367–73.
Matthies, Melanie L., Mario Svirsky, Joseph Perkell, and Harlan Lane. 1996. Acoustic and Articulatory Measures of Sibilant Production with and without Auditory Feedback from a Cochlear Implant. *Journal of Speech and Hearing Research* 39(5): 936–46.
McClean, Michael. 1977. Effects of auditory masking on lip movement during speech. *Journal of Speech and Hearing Research* 20: 731–741.
McKay, Collette, and Katherine R. Henshall. 2002. Frequency-to-electrode allocation and speech perception with cochlear implants. *Journal of the Acoustical Society of America* 111(2): 1036–1044.
Medical Electronics. 1999. *Information for Candidates: A Guide to the COMBI 40+ Cochlear Implant System.*
Mendoza, Monica. 1998. Viewers love the realism // Educator for deaf assists 'ER' with characters, story. *The Arizona Daily Star*, December 6, 1998. p. 1.B.
Osberger, Mary Joe, Monica Maso, and Leslie K. Sam. 1993. Speech Intelligibility of Children with Cochlear Implants, Tactile Aids, or Hearing Aids. *Journal of Speech and Hearing Research* 36: 186–203.
Parkinson, Aaron J., Wendy S. Parkinson, Richard S. Tyler, Mary W. Lowder, and Bruce J. Gantz. 1998. Speech Perception Performance in Experienced

Cochlear-Implant Patients Receiving the SPEAK Processing Strategy in the Nucleus Spectra-22 Cochlear Implant. *Journal of Speech, Language, and Hearing Research* 41: 1073–1087.

Perkell, Joseph, Harlan Lane, Mario Svirsky, and Jane Webster. 1992. Speech of Cochlear Implant Patients: A Longitudinal Study of Vowel Production. *Journal of the Acoustical Society of America* 91(5): 2961–78.

Peterson, Gordon E. and Harold L. Barney. 1952. Control Methods Used in a Study of the Vowels. *Journal of the Acoustical Society of America* 24: 175–184.

Platinum Series Implant. 2000. Advanced Bionics Corporation. 20 January 2000 <http://www.cochlearimplant.com/plat_SP.html>.

Richardson, Louise M., Peter A. Busby, Peter J. Blamey, and Graeme M. Clark. 1998. Studies of prosody perception by cochlear implant patients. *Audiology.* 37(4): 231–245.

Rubin-Spitz, Judith and Nancy S. McGarr. 1990. Perception of Terminal Fall Contours in Speech Produced by Deaf Persons. *Journal of Speech and Hearing Research* 33(1): 174–180.

Seligman, P. and McDermott, H. Architecture of the Spectra 22 speech processor. 1995. *Annals of Otology, Rhinology, and Laryngology.* 104(9) part 2: 139–141.

Siegel, Gerald M. and Herbert L. Pick Jr. 1974. Auditory feedback in the regulation of voice. *Journal of the Acoustical Society of America* 56(5): 1618–1624.

Skinner, Margaret W., Graeme M. Clark, Lesley A. Whitford, Peter M. Seligman, Steven J. Staller, David B. Shipp, Jon K. Shallop, Colleen Everingham, Christine M. Menapace, Patti L. Arndt, Trisha Antogenelli, Judith A. Brimacombe, Sipke Pijl, Paulette Daniels, Catherine R. George, Hugh J. Mcdermott, and Anne L. Beiter. 1994. Evaluation of a new spectral peak coding strategy for the Nucleus 22 channel cochlear implant system. *American Journal of Otology.* 15 suppl. 2: 15–27.

Superhuman Bodies. 2001. BBC. 19 April 2001 <http://www.bbc.co.uk/science/programmes/superhuman.shtml>.

Svirsky, Mario A., Harlan Lane, Joseph S. Perkell, and Jane Wozniak. 1992. Effects of short-term auditory deprivation on speech production in adult cochlear implant users. *Journal of the Acoustical Society of America* 92(3): 1284–1300.

Svirsky, Mario A., Alicia Silveira, Hamlet Suarez, Heidi Neuberger, Ted T. Lai and Peter M. Simmons. 2001. Auditory Learning and Adaptation after Cochlear Implantation: A Preliminary Study of Discrimination and Labeling of Vowel Sounds by Cochlear Implant Users. *Acta Otolaryngol.* 121: 262–265.

Tye-Murray, Nancy. 1992. Young cochlear implant users' response to delayed auditory feedback. *Journal of the Acoustical Society of America* 91(6): 3483–3486.

Tye-Murray, Nancy and Karen Iler Kirk. 1993. Vowel and Diphthong Production by Young Users of Cochlear Implants, and the Relationship between the Phonetic Level Evaluation and Spontaneous Speech. *Journal of Speech and Hearing Research* 36(3): 488–502.

Tye-Murray, Nancy, Linda Spencer, Elizabeth Gilbert Bedia, and George Woodworth. 1996. Differences in Children's Sound Production When Speaking with a Cochlear Implant Turned On and Turned Off. *Journal of Speech and Hearing Research* 39(3): 604–10.

Tye-Murray, Nancy, Linda Spencer, and Elizabeth Gilbert-Bedia,. 1995. Relationships between speech production and speech perception in young cochlear-implant users. *Journal of the Acoustical Society of America* 98(5): 2454–60.

Tye-Murray, Nancy, Linda Spencer, and George G. Woodworth. 1995. Acquisition of Speech by Children Who Have Prolonged Cochlear Implant Experience. *Journal of Speech and Hearing Research* 38(2): 327–37.

Tyler, Richard S., and Brian C.J. Moore. 1992. Consonant Recognition by Some of the Better Cochlear-Implant Patients. *Journal of the Acoustical Society of America* 92(6): 3068–77.

Välimaa, Taina T., Taisto K. Määttä, Heikki J. Löppönen, and Martti J. Sorri. 2002(1). Phoneme Recognitions and Confusions with Multichannel Cochlear Implants: Vowels. *Journal of Speech, Language, and Hearing Research* 45: 1039–1054.

Välimaa, Taina T., Taisto K. Määttä, Heikki J. Löppönen, and Martti J. Sorri. 2002(2). Phoneme Recognitions and Confusions with Multichannel Cochlear Implants: Consonants. *Journal of Speech, Language, and Hearing Research* 45: 1055–1069.

Vandali, Andrew E. 2001. Emphasis of short-duration acoustic speech cues for cochlear implant users. *Journal of the Acoustical Society of America* 109(5): 2049–2061.

van Hoesel, R.J.M., Y.C. Tong, R.D. Hollow, and G.M. Clark. 1993. Psychophysical and Speech Perception Studies: A Case Report on a Binaural Cochlear Implant Subject. *Journal of the Acoustical Society of America* 94(6): 3178–89.

Van Summers, W., David B. Pisoni, Robert H. Bernacki, Robert I. Pedlow, and Michael A. Stokes. 1988. Effects of noise on speech production: Acoustic and perceptual analyses. *Journal of the Acoustical Society of America* 84(3): 917–928.

Vick, Jennell C., Harlan Lane, Joseph S. Perkell, Melanie L. Matthies, John Gould, and Majid Zandipour. 2001. Covariation of cochlear implant users' perception and production of vowel contrasts and their identification by listeners with normal hearing. *Journal of Speech, Language, and Hearing Research* 44: 1257–1267.

Waldstein, Robin S. 1990. Effects of Postlingual Deafness on Speech Production: Implications for the Role of Auditory Feedback. *Journal of the Acoustical Society of America* 88(5): 2099–2114.

Waltzman, Susan B. and Cohen, Noel L. 1999. Implantation of Patients with Prelingual Long-Term Deafness. *Annals of Otology, Rhinology, and Laryngology* 108: 84–87.

Wiesner, David. 1991. *Tuesday*. New York: Clarion Books.

Wilson, Blake S., Dewey T. Lawson, Charles C. Finley, and Robert D. Wolford. 1993. Importance of Patient and Processor Variables in Determining Outcomes With Cochlear Implants. *Journal of Speech and Hearing Research* 36: 373–379.

Yost, William A. 2000. *Fundamentals of Hearing: An Introduction*. San Diego: Academic Press.

Zeng, Fan-Gang, Ginger Grant, John Niparko, John Galvin, Robert Shannon, Jane Opie, and Phil Segel. 2001. Speech dynamic range and its effect on cochlear implant performance. *Journal of the Acoustical Society of America* 111(1): 377–386.

Ziese, M., A. Stützel, H. von Specht, K. Begall, B. Freigang, S. Sroka, and P. Nopp. 2000. Speech Understanding with the CIS and the n-of-m Strategy in the MED-EL COMBI 40+ System. *Annals of Otology, Rhinology, and Laryngology*. 62: 321–329.

Zimmerman-Phillips, Susan and Murad, Carol. 1999. Programming Features of the Clarion® Multi-Strategy™ Cochlear Implant. *Annals of Otology, Rhinology, and Laryngology*. 108(4) part 2: 17–21

Index

Lightning Source UK Ltd.
Milton Keynes UK
27 May 2010

154836UK00002B/26/P

9 780415 976046